A Manual of Cooperative Learning Worksheets to accompany Masterton and Hurley's Chemistry: Principles and Reactions

Fifth Edition

Cecile N. Hurley

University of Connecticut

THOMSON
— ✳ —™
BROOKS/COLE

Australia • Canada • Mexico • Singapore • Spain • United Kingdom • United States

Printed in the United States of America
2 3 4 5 6 7 07 06

Printer: Globus Printing Company, Inc.

ISBN: 0-534-40879-6

For more information about our products, contact us at:
Thomson Learning Academic Resource Center
1-800-423-0563

For permission to use material from this text,
contact us by:
Phone: 1-800-730-2214
Fax: 1-800-731-2215
Web: http://www.thomsonrights.com

Brooks/Cole—Thomson Learning
10 Davis Drive
Belmont, CA 94002-3098
USA

Asia
Thomson Learning
5 Shenton Way #01-01
UIC Building
Singapore 068808

Australia/New Zealand
Thomson Learning
102 Dodds Street
Southbank, Victoria 3006
Australia

Canada
Nelson
1120 Birchmount Road
Toronto, Ontario M1K 5G4
Canada

Europe/Middle East/South Africa
Thomson Learning
High Holborn House
50/51 Bedford Row
London WC1R 4LR
United Kingdom

Latin America
Thomson Learning
Seneca, 53
Colonia Polanco
11560 Mexico D.F.
Mexico

Spain/Portugal
Paraninfo
Calle/Magallanes, 25
28015 Madrid, Spain

Preface

Collaborative (or, cooperative) learning is not new. Most general chemistry students bring ample experience with it from their elementary and high-school careers. What *is* new is its growing use in colleges and universities as a means of promoting improved learning. Cooperative learning does that in several ways, including

- providing a new, more active structure for weekly discussion sections

- giving students the opportunity to verbalize their thought processes clearly enough to make them understandable to their group peers

- exposing students to new and different ways that their peers may use to solve problems, visualize concepts, and think about how to apply theory

- encouraging students to work together on problems that promote deeper understanding than routine homework questions.

Studies at the University of Connecticut show that students in sections with collaborative learning have averaged half a letter grade higher than those in standard sections of the same course. This approach has been particularly effective for women, spectacularly so for minority women. Consequently, all sections of our general chemistry course now use group learning in the weekly discussion period. The results of our study (which presents a detailed statistical analysis of the experinece with collaborative learning at the University of Connecticut) can be found in the article "Cooperative Learning in the Quiz Section in General Chemistry" which appears in *Proc. Frontiers in Education 23,* Washington, DC, IEEE & ASEE, 1993, 162-166.

The focus of our cooperative learning is the weekly worksheet. This manual has three different ones for Chapters 1–19 of the text. (Chapters 20–22 are largely descriptive and use the principles discussed in earlier chapters. There are no worksheets for these three chapters.) The section *Organization* discusses various approaches to breaking a class up into such groups. Teams of 3-to-5 students discuss the worksheet each week under the supervision of an instructor or teaching assistant.

If students get "stuck" or ask for clarification of a point, the instructor provides a hint or asks leading questions to get them moving again. It is *very* rare for them to coach students actively on how to analyze a problem, and rarer still for them to present a step-by-step solution. Instead, when the groups complete an item, they volunteer an answer and ask the instructor to appraise the reasoning that led them to it.

The worksheets are all of comparable difficulty. Each one is roughly half conceptual and half quantitative, with the latter part more challenging than the assigned homework problems the students work out before they come to discussion. It should take the student teams roughly an hour to do the entire worksheet. Our experience has led us to divide the worksheets into 3 or 4 parts, assigning a part to each group of 3 to 5 students. The groups work on their assignignments for about 40 minutes and spend the rest of the hour presenting their solutions and answers to ther assigned question to the rest of the class.

The conceptual half of each worksheet aims to teach students to be precise in both their thinking and their language. They learn that words such as *never* and *always* are rarely valid in chemistry, and that sweeping generalizations are risky. They teach each other that decreasing *by* 25% is not the same as decreasing *to* 25%. It is in this area of chemical reasoning that we have seen the students make the greatest improvement. The class mean on the conceptual part of the hour exams used to hover in the range of 50-to-60%. After we switched to group learning during the quiz sections, that climbed dramatically — to a range of 68-to-75%.

These worksheets have been used in other ways by instructors at other universities. Some use it as the framework for a quiz section where the whole class participates (as a group) in discussing the entire worksheet with the TA (or instructor) as group leader. If the format of your quiz section is such that a TA is responsible for the activity that takes place in his/her discussion section, then these worksheets are particularly useful to the new TA's who have no experience in selecting the type of questions that should be discussed.

Other instructors assign the worksheet as a group take-home exam. Still others assign them for extra-credit work or as a make-up for missed assignments. However you choose to use them, it is our experience that a certain amount of time should be spent listening to the student's reasoning on how he/she arrived at the answers handed in. These worksheets were never intended to substitute for written homework problems. Their efficacy seems to be largely due to having the students verbalize their reasoning process.

It is a pleasure to acknowledge the contributions of individuals and institutions to this endeavor. The development of the collaborative approach to general chemistry, which stimulated the creation of these worksheets, was supported in part by a grant from the National Science Foundation Curriculum Development Program in Chemistry (USE – 9155980). That grant also provided partial support for the testing and assessment of the efficacy of the new instructional mode. Joel Tolentino and Maria Cecilia de Mesa reviewed these worksheets for accuracy. James F. Hurley served as sounding board, critic, and booster. While the deficiencies are my sole responsibility, the production of these worksheets in final form owes much to their efforts.

Cooperative learning is an exciting tool that has shown it can be of real value in general chemistry. It can help instructors as well as students, and I hope that these worksheets can help you implement a successful program of group work in your class.

– Cecile N. Hurley
Storrs, Connecticut
May, 2003

Organization

The organization of a class or quiz section into groups is determined in large part by the nature of the course, the number of students to be divided up, and the role that collaborative learning will play. What follows is a brief description of the different ways that we have used to organize the students in our discussion sections.

- Random assignment
  Students are assigned numbers and groups are formed by picking numbers from a bag.

- Peer choice
  The students tell about themselves at the first meeting. They are then asked to group themselves. Most often the groups are formed by people with the same major or people in the same or nearby dorms.

- Group by gender
  The teaching assistant assigns the students to single gender groups, and announces the members of each group at the first meeting.

- Group by ability
  After the first exam, the students are grouped according to grade. Before that, the groups are set up by using SAT math scores.

At the University of Connecticut, we have tried all four groupings. We have tried both keeping the same grouping throughout the entire semester, and reconstituting the groups every 2–3 weeks. In general, the latter practice seems to have worked better. Students like the fact that they can get to know more of the students in their section, and if some personalities don't get along, relief is at most two weeks away.

Grouping by peer choice seemed the least desirable method of organization. Since most of the students are freshmen, they don't know each other that well and it can be awkward, even embarrassing, if someone does not get picked.

Group dynamics favor grouping by gender. Women seem more adept at working as a team. The men are less inclined to group effort, but generally do just as well after a week or two.

We have gotten the best results by grouping according to ability. A C student in a group of A students feels intimidated, or relies totally on the other students. Grouped together, A and B students get more of the worksheet done during class time. They require less supervision from the teaching assistant and free the instructor to spend more time with the slower students. The C, D, and F students are less afraid to speak up when they are grouped together. They more readily share their insecurities and the parts of the material that they do not understand. It is also a tremendous boost to their self-esteem to get correct answers on their own.

If you would like more information about group organization, I can be reached by telephone, (860) 486–3795, or by e–mail at *cecile.hurley@uconn.edu*

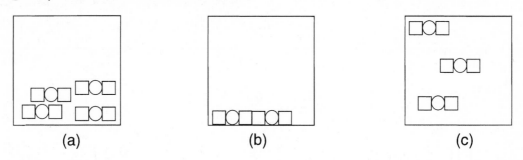

Matter and Measurements

Worksheet A

A. Answer the following questions.

 1. For each of the sentences below, write E if the italicized number is exact and U if it has experimental uncertainty.

  _____ a. There are *12* eggs in a dozen.

  _____ b. The baby weighs *27.6* lbs.

  _____ c. The grocer threw away *17* rotten oranges.

  _____ d. There were *12,000* people at the protest rally.

  _____ e. There are *5280* ft in a mile.

 2. Label each of the properties of iodine as intensive (I) or extensive (X).

  _____ a. Its density is 4.93 g/mL.

  _____ b. It is purple.

  _____ c. It melts at 114°C.

  _____ d. One hundred grams of melted iodine has a volume of 122 mL.

  _____ e. Some crystals are large.

 3. Consider the compound X_2Y represented as ▢◯▢ . Indicate the physical state of X_2Y represented in the boxes below.

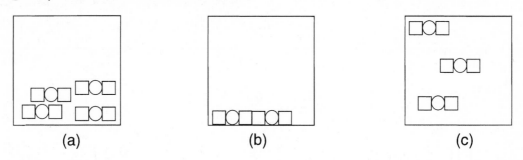

     (a)   (b)   (c)

 4. How many revolutions will the second hand of a clock make in five days? (Solve by conversion factor method.)

B. Label each statement as true or false. If the statement is false, make it true.

_____ 1. The Kelvin degree is the same as the Celsius degree.

_____ 2. The correct answer to $\dfrac{13.71 \times 10^{18}}{121.5 + (1.32 \times 10^3) - 12}$ is 9.6×10^{16}.

_____ 3. The solubility of compounds in water can increase or decrease with increasing temperature.

_____ 4. A homogeneous mixture can always be separated into its components by distillation.

_____ 5. If a mixture can be separated into its components by filtration, then it must be a heterogeneous mixture.

C. A dilute solution of syrup and water is prepared by mixing 45.00 mL of water (d = 0.9982 g/mL) with 75.00 mL of syrup (d = 1.124 g/mL).
1.   What is the percent by mass of syrup in the diluted solution?

2.   If the total volume of the dilute solution is 112.73 mL, what is the density of the resulting solution?

D. A compound X has a solubility of 38 g/100 g H_2O at 75°C. A solution made up of 15 g of X and 43 g of water is prepared at 75°C. When the solution is cooled to room temperature (25°C), four grams of X crystallize out of solution. What is the solubility of X (g X/100 g H_2O) at room temperature?

Matter and Measurements

Worksheet B

A. Answer the following questions:
1. How many significant figures are there in

   a. 12.060 g
   b. 150 mL
   c. 0.003020 km

2. Five liquids have the following densities in g/cm^3:

   (1)   0.445
   (2)   0.927
   (3)   1.110
   (4)   3.643
   (5) 12.66

   You are given 10.0 g samples of each liquid. Which one has the largest volume?

B. Calculate

   1. $\dfrac{6.24 + 1374}{2.861}$

   2. $\dfrac{32.1 - 0.35}{9871 + 0.2}$

   3. $\dfrac{34.3 \times 10^{23}}{0.0071} + 124$

C. On an alien planet, two different scales are used. Two quantities in both scales are given below:

   27.3 zigs = 1 zag
   15.2 quarks^2 = 2.2 zirks

   Convert 398 zirks/(zig)^2 to quarks/zag

D. Given the following solubility curves, answer the following questions:

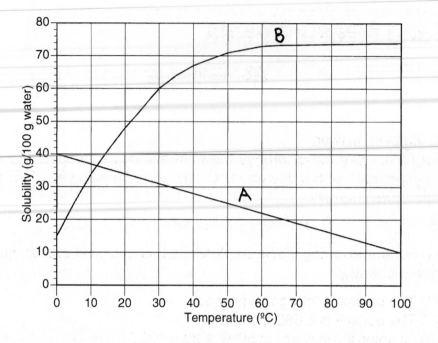

1. In which of the two compounds can more solute be dissolved in the same amount of water when the temperature is increased?

2. At what temperature is the solubility of both compounds the same?

3. Consider Compound A.
    a. Describe the effect that an increase in temperature has on the solubility of the compound.

    b. How many grams of solid can be dissolved in 42 grams of water at 40°C?

    c. A student adds 25 g of the compound to 85 g of water at 20°C. Will it all dissolve? The solution is then heated to 90°C. How much, if any, of the solute will crystallize out?

4. Consider Compound B.
    a. A student tries to dissolve 65 g of the compound in 100 g of water at 30°C. Will it all dissolve?

    b. If not, how much will remain undissolved?

    c. Describe the effect that an increase in temperature has on the solubility of this compound.

Matter and Measurements

Worksheet C

A. Choose the best answer.

1. On a recent day the difference between the lowest temperature and the highest temperature of the day was 9°C. By how many degrees Fahrenheit did the temperature change?

   A. 5   B. 9   C. 16   D. 37   E. 48

2. Which of the following would be described as chemical property(ies) of sodium periodate, $NaIO_4$?

   (1) The crystal is tetragonal in shape.
   (2) The density is 3.865 g/cm^3 at 16°C.
   (3) Its solubility in water at 50°C is 37 g/100.00 g H_2O.
   (4) It reacts with ethylene glycol.

   A. (1),(2)   B. (2),(4)   C. (1),(3)   D. (4)   E. (1),(2),(3),(4)

3. The number of significant figures in the answer to 156.0 g – 121.02 g is

   A. 2   B. 3   C. 4   D. 5

4. Five liquids have the following densities in g/cm^3:

   (1) 1.62   (2) 0.912   (3) 3.49   (4) 2.08   (5) 0.982

   You are given samples of equal mass of each liquid. Which one has the largest volume?

   A. (1)   B. (2)   C. (3)   D. (4)   E. (5)

5. The correct answer to $\dfrac{2.634\,g \times 12.0\,g}{3.15\,g} + 4.2\,g$ is

   A. 14.23 g   B. 14.2 g   C. 14.234 g   D. 14 g

B. A cube of gold, 1.30 cm on a side, is added to a pycnometer of unknown volume. Liquid acetone is used as the fluid. The following data are collected:

mass of empty flask	121.06 g
mass of flask + gold	163.46 g
mass of flask + acetone	191.50 g
mass of flask + acetone + gold	232.17 g
density of acetone	0.787 g/cm^3

1. What is the volume of the flask?

2. What is the density of gold?

C. Given the following solubility graph for a compound M:

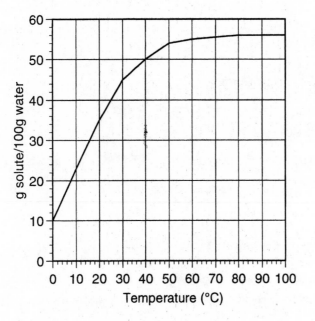

1. How many grams of the solid can be dissolved in 25 g of water at 40°C?

2. Assuming that at 40°C, the volume of the solution is equal to the volume of the water, what is the density of the solution formed above? ($d_{H_2O} = 1.00$ g/mL)

3. A student adds 35 g of the substance to 72 g of water at 25°C. Will it all dissolve? If not, how much remains undissolved?

Matter and Measurements

Answers

Worksheet A

A. 1. a. E   b. U   c. E   d. U   e. E

    2. a. I   b. I   c. I   d. X   e. X

    3. a. liquid   b. solid   c. gas

    4. 7.2×10^3 revolutions

B. 1. True

    2. False — replace 9.6×10^{16} with 9.59×10^{15}

    3. True

    4. False — replace always with sometimes

    5. True

C. 1. 65.24%   2. 1.1463 g/mL

D. 26 g/100 g H_2O

Worksheet B

A. 1. a. 5   b. ambiguous (2 or 3)   c. 4

    2. (1)

B. 1. 482.4   2. 0.00322   3. 4.8×10^{26}

C. 1.4×10^3 quarks/zag

D. 1. B

    2. $\approx 12°C$

    3. a. Increasing the temperature decreases the solubility

      b. ≈ 12 g   c. yes; ≈ 13 g

    4. a. No   b. ≈ 5 g

      c. Increasing the temperature increases the solubility to a point ($\approx 60°C$).
After that an increase in temperature no longer has an effect.

Worksheet C

A. 1. C   2. D   3. B   4. B   5. B

B. 1. 89.5 cm³   2. 19.3 g/cm³ or 19 g/cm³ depending on method of calculation

C. 1. ≈ 12 g   2. ≈ 1.5 g/mL   3. no; ≈ 6 g

2

Atoms, Molecules, and Ions

Worksheet A

A. Which of these statements resulting from Dalton's atomic theory are still true today?
   1. All atoms of an element are identical in mass.
   2. Atoms are indivisible and indestructible.
   3. Atoms of elements combine in the ratios of small whole numbers to form compounds.

B. Write the atomic symbol for the element
   1. that has a total of 60 protons, neutrons, and electrons with equal numbers of all three.
   2. whose ion has the following properties: a +2 charge, 10% more protons than electrons, 26 neutrons.
   3. whose mass number is 120 and has 40% more neutrons than protons.

C. True or False
   1. Isotopes of the same element have the same mass number.
   2. It is impossible for phosphorus and sulfur to be isotopic to each other.
   3. It is possible for phosphorus and sulfur to have the same number of electrons.
   4. The number of protons must always be equal to or greater than the number of electrons.

D. Name the following:

   1. BrI_3 _____

   2. Ca_3N_2 _____

   3. $Fe_2(Cr_2O_7)_3$ _____

   4. XeF_6 _____

   5. N_2O_4 _____

E. Write the formula for the following:

1. copper(II) phosphate _____

2. phosphorus trichloride _____

3. potassium sulfite _____

4. iron(III) nitrate _____

5. barium perchlorate _____

6. ammonia _____

Atoms, Molecules, and Ions

Worksheet B

A. Circle the correct statement(s).
  1. The electron
     a. is positively charged.
     b. is outside the nucleus.
     c. is part of the nucleus.
     d. is always equal in number to protons.
     e. has all the properties of the element.

  2. The proton
     a. is positively charged.
     b. is outside the nucleus.
     c. is part of the nucleus.
     d. is always equal in number to electrons.
     e. has the same properties for all elements.

  3. Neutrons
     a. are uncharged.
     b. have no mass.
     c. are located in a small space at the center of the atom.
     d. are always twice the number of protons.
     e. determine the atomic number of the element.

  4. NH_4Cl
     a. is made up of 2 ions.
     b. is a molecule.
     c. is made up of nitrogen, hydrogen and chlorine molecules.
     d. has 28 protons in one formula unit.
     e. has a cation and an anion.

  5. Uncharged isotopes always have the same
     a. mass number – A.
     b. atomic number – Z.
     c. number of electrons.
     d. number of neutrons.
     e. spot in the periodic table.

6. $^{19}F^-$, ^{20}Ne, and $^{24}Mg^{2+}$ all have the same
    a. mass number.   b. atomic number.
    c. number of neutrons.   d. number of electrons.
    e. period in the periodic table.

7. If an element is positively charged, then the atom has
    a. gained electrons.   b. lost electrons.
    c. gained neutrons.   d. lost neutrons.
    e. gained protons.   f. lost protons.

B. Name the following:

1. $BaCO_3$ _____

2. CoN _____

3. ICl_3 _____

4. SF_6 _____

5. $Fe(ClO_3)_2$ _____

C. Write the formula for the following:

1. sulfur dioxide _____

2. ammonium permanganate _____

3. nickel(III) sulfate _____

4. lithium hypobromite _____

5. tetraphosphorous decaoxide _____

6. sulfuric acid _____

D. Write the atomic symbol for the element
1. that has a total of 90 protons, neutrons, and electrons with equal numbers of all three.

2. that has a mass number of 184 and has 36 more neutrons than protons.

3. that has 20 more neutrons than electrons, has a mass number of 126, and whose ion has a −2 charge.

Atoms, Molecules, and Ions

Worksheet C

A. Answer the following questions.

1. Scientists are trying to synthesize elements with more than 106 protons. State the expected atomic number of
   a. the newest inert gas.
   b. the new element with properties similar to those of copper, silver and gold.
   c. the new element that will behave like the halogens.
   d. the new metal whose ion will have a +2 charge and is not a transition element.
   e. the new element that will start Period 8.

2. Twenty five grams of a red compound are sealed into a flask that contains only 12.0 g of oxygen. After reaction, the flask is opened, and found to contain 23.0 g of an insoluble yellow solid, and 10.0 g of liquid. How much oxygen was unused?

3. Consider the elements arsenic, manganese, selenium, and sodium. Among these elements, name the element(s) that
   a. has(have) chemical properties most like potassium.
   b. is/are non-metal(s).
   c. is/are transition metal(s).
   d. is/are in group 16.
   e. is/are metalloid(s).
   f. has/have an ion with 23 electrons.
   g. has an oxoanion called permanganate. Write the formula for the permanganate ion. What is a reasonable guess for the formula of the manganate ion?

4. A student saw the following nuclear symbol for an unknown element: $^{14}_{6}X$. Which of the following statements about X and $^{14}_{6}X$ are true?
   a. X is carbon.
   b. X is silicon.
   c. X is oxygen.
   d. X has 6 neutrons in its nucleus.
   e. Electrically neutral X has 8 electrons.
   f. $^{12}_{6}X$ is more stable than $^{14}_{6}X$.

B. Name the following:

1. NH_3 ——————————————————————————

2. $Ni_2(Cr_2O_7)_3$ ——————————————————————

3. $(NH_4)_2CO_3$ ——————————————————————

4. P_4O_{10} ————————————————————————

5. Ba_3N_2 ————————————————————————

6. $HClO_3$ (aq) ————————————————————————

7. HI (aq) ————————————————————————

8. HBr (g) ————————————————————————

C. Write the formula for the following:

1. methane ————————————————————————

2. dinitrogen oxide ——————————————————————

3. iron(III) nitrite ————————————————————————

4. xenon tetrachloride ——————————————————————

5. copper(II) hydroxide ——————————————————————

6. aluminum sulfide ——————————————————————

Atoms, Molecules, and Ions

Answers

Worksheet A

A. 3

B. 1. $^{40}_{20}Ca$ 2. $^{48}_{22}Ti$ 3. $^{120}_{50}Sn$

C. 1. F   2. T   3. T   4. F

D. 1. bromine triiodide   2. calcium nitride
   3. iron(III) dichromate   4. xenon hexafluoride
   5. dinitrogen tetraoxide

E. 1. $Cu_3(PO_4)_2$ 2. PCl_3 3. K_2SO_3 4. $Fe(NO_3)_3$
   5. $Ba(ClO_4)_2$ 6. NH_3

Worksheet B

A. 1. b   2. a, c, e   3. a, c   4. a, d, e
   5. b, c, e   6. d   7. b

B. 1. barium carbonate   2. cobalt(III) nitride
   3. iodine trichloride   4. sulfur hexafluoride
   5. iron(II) chlorate

C. 1. SO_2 2. NH_4MnO_4 3. $Ni_2(SO_4)_3$ 4. LiBrO
   5. P_4O_{10} 6. H_2SO_4

D. 1. $^{60}_{30}Zn$ 2. $^{184}_{74}W$ 3. $^{126}_{52}Te$

Worksheet C

A. 1. a. 118   b. 111   c. 117   d. 120   e. 119
   2. 4.0 g
   3. a. Na   b. Se   c. Mn   d. Se
      e. As   f. Mn   g. Mn; MnO_4^-; MnO_3^-
   4. a, f

B. 1. ammonia   2. nickel(III) dichromate
   3. ammonium carbonate   4. tetraphosphorous decaoxide
   5. barium nitride   6. chloric acid
   7. hydroiodic acid   8. hydrogen bromide

C. 1. CH_4 2. N_2O 3. $Fe(NO_2)_3$ 4. $XeCl_4$ 5. $Cu(OH)_2$
   6. Al_2S_3

Mass Relations in Chemistry; Stoichiometry

Worksheet A

A. Given the following mass spectrum for an element, estimate the atomic mass of a sample of the naturally occurring element.

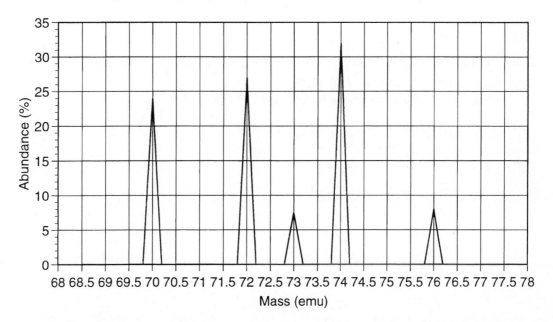

B. If one were to cover the area of the 48 contiguous states with a mole of M&M's, how thick would the layer be? Assume the area being covered is entirely flat.

Helpful information and/or materials (maybe):
88 M&M's fill a 100.0 mL cylinder   5280 ft = 1 mile
area of 48 states = 3.02×10^6 mi^2 Avogadro's No. = 6.022×10^{23}
1 mL = 1 cm^3 2.54 cm = 1 in

C. Which of the following statements is/are true? If the statement is false, make it true.
A 558.5 gram sample of iron contains
1. 100.0 mol of iron.
2. 6.022×10^{23} atoms of iron.
3. ten times as many atoms as 0.5200 grams of chromium.
4. twice as many atoms as 60.06 grams of carbon.

D. Box A contains 5 molecules of nitrogen (N_2) and 15 molecules of hydrogen (H_2). Box B contains 10 molecules of NH_3.

    1. Make a pictorial representation of Box A and Box B using open circles to represent N atoms and squares to represent H atoms.

    2. Compare box A and box B with respect to
        a. number of atoms of N and H.
        b. number of discrete particles.
        c. mass.
        (Do not use your calculator. Use only logical reasoning.)

    3. a. If the contents of Box A are the reactants and those of box B the products, write a reaction to represent boxes A and B.
        b. Reduce the coefficients of the reaction in (a) to smallest whole number coefficients.

E. A sample contains only carbon and hydrogen atoms. It was burned in oxygen producing 18.48 g of CO_2 and 6.30 g of H_2O.

    1. What is the simplest formula of the compound?

    2. How many grams of sample were burned?

Mass Relations in Chemistry; Stoichiometry

Worksheet B

A. Which of the following statements is/are true? If the statement is false, make it true.
1. The compound $C_6H_{12}O_2N$
   a. has the simplest formula $C_3H_6ON_{\frac{1}{2}}$.
   b. contains 21 atoms per mole.
   c. has twice as many atoms of carbon as hydrogen.
   d. has three times as many moles of carbon as moles of oxygen atoms.
   e. has half as many grams of hydrogen as carbon.

2. A 6.539 g sample of zinc has
   a. 65.39 moles of zinc.
   b. 6.022×10^{23} atoms of zinc.
   c. a hundred times as many atoms as 0.197 g of gold (Au).
   d. twice as many moles as 13.08 g of sulfur.

B. Iron is important in the body primarily because it is present in red blood cells and acts to carry oxygen to the various organs. There are about 2.6×10^{13} red blood cells in all of the blood of an adult human. All of adult human blood contains a total of 2.9 g of iron.
1. How many moles of iron are present in all of the blood of an adult human?

2. How many iron atoms are there in each blood cell?

C. A compound is made up of carbon, hydrogen, and oxygen atoms. When burned in oxygen, 3.584 g of CO_2 and 1.957 g of H_2O are formed. The mass of oxygen in the compound is 36.3% that of the CO_2 produced.
1. How much sample was burned?

2. What is the empirical formula of the compound?

D. Chromium has the following isotopic mass and abundances (%):

Isotope	Atomic Mass (amu)	Abundance (%)
Cr–50	49.9461	4.35
Cr–52	51.9405	83.79
Cr–53	52.9407	9.50
Cr–54	53.9389	2.36

1. Sketch the mass spectrum for chromium.

2. Estimate the atomic mass of a sample of naturally occurring chromium.

Mass Relations in Chemistry; Stoichiometry

Worksheet C

A. Label each statement as true (T) or false (F).

_____ 1. The absolute masses of atoms have the unit *mole*.

_____ 2. There are 6.022×10^{23} atoms of sodium in 22.99 lbs of sodium.

_____ 3. In N_2O_4, the mass of oxygen is twice that of nitrogen.

_____ 4. One mole of chlorine atoms has a mass of 35.45 g.

_____ 5. Boron has an average mass of 10.81 amu and has two isotopes, B–10 (10.01 amu) and B–11 (11.01 amu).
There is more naturally occurring B–10 than B–11.

B. Suppose that S–32 ($^{32}_{16}$S) was taken as the standard for expressing atomic masses and assigned an atomic mass of 10.00 amu. Estimate the formula mass of copper(I) oxide.

C. A compound is made up of carbon, hydrogen, and oxygen atoms. It has 52.14% C and 34.73% O.
1. What is the simplest formula of the compound?

2. How many grams of water are obtained when 1.000 g of the compound is burned in oxygen?

D. Magnesium has three isotopes (Mg–24, Mg–25, and Mg–26). Complete the mass
    spectrum for magnesium given below.

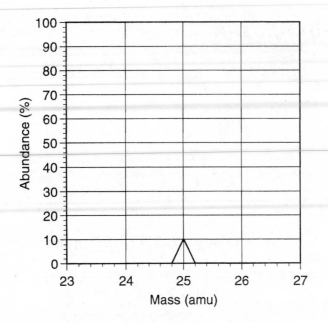

E. Box A contains 24 atoms of arsenic (As) and 18 molecules of oxygen (O_2). Box B
    contains 12 molecules of As_2O_3.
    1. Compare box A and box B with respect to
        a. number of atoms of As and O.
        b. number of discrete particles.
        c. mass.
            (Do not use your calculator. Use only logical reasoning.)

    2. a. If the contents of Box A are the reactants and those of box B the products,
           write a reaction to represent boxes A and B.
        b. Reduce the coefficients of the reaction in (a) to smallest whole number coef-
           ficients.

Mass Relations in Chemistry; Stoichiometry

Answers

Worksheet A

A. $\approx$ 72 amu

B. 54 mi

C. 1. False – <u>10.00</u> mol of iron   2. False – <u>6.022×10^{24}</u> atoms
    3. False – <u>1000</u> times   4. True

D. 1.   Box A   Box B

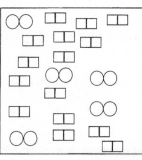 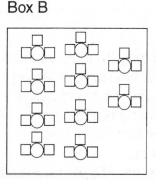

    2.   a.   same number of N atoms in Boxes A and B (10)
             same number of H atoms in Boxes A and B (30)
       b.   Different number of discrete particles (20 in Box A; 10 in Box B)
       c.   mass of Box A = mass of Box B
    3.   a. $5\,N_2 + 15\,H_2 \rightarrow 10\,NH_3$
       b. $N_2 + 3\,H_2 \rightarrow 2\,NH_3$

E. 1. C_3H_5 2. 5.743 g

Worksheet B

A. 1.   a.   False – <u>$C_6H_{12}O_2N$</u>   b.   False – per <u>molecule</u>
       c.   False – <u>half</u> as many   d.   True
       e.   False – <u>one – sixth</u> times
    2.   a.   False – <u>0.1000</u> moles   b.   False – <u>6.022×10^{22}</u> atoms of zinc
       c.   True   d.   False – a <u>fourth</u> as many

B. 1.   0.052 mol   2. 1.2×10^9

C. 1.   2.50 g   2. $C_3H_8O_3$

D. 1.

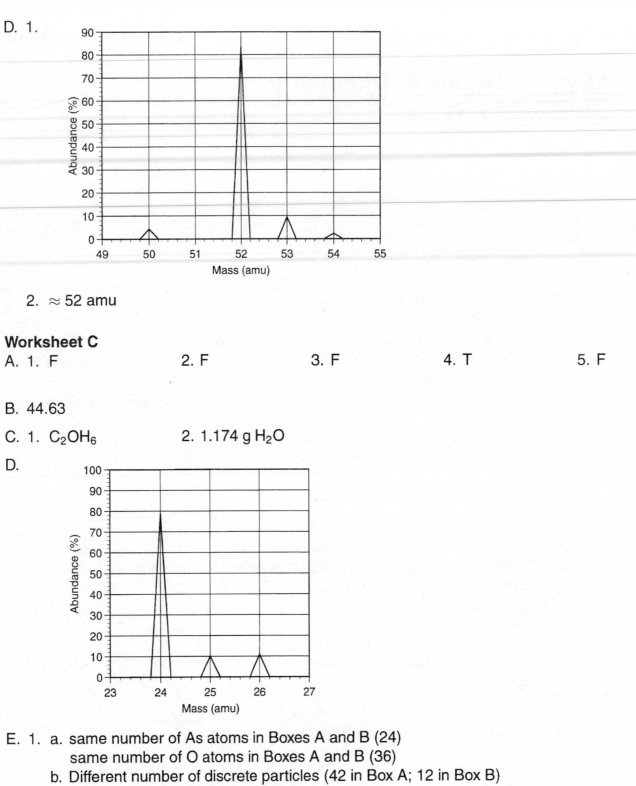

2. ≈ 52 amu

Worksheet C

A. 1. F   2. F   3. F   4. T   5. F

B. 44.63

C. 1. C_2OH_6 2. 1.174 g H_2O

D.

E. 1. a. same number of As atoms in Boxes A and B (24)
            same number of O atoms in Boxes A and B (36)
       b. Different number of discrete particles (42 in Box A; 12 in Box B)
       c. mass of Box A = mass of Box B
    2. a. $24\,As + 18\,O_2 \rightarrow 12\,As_2O_3$
        b. $4\,As + 3\,O_2 \rightarrow 2\,As_2O_3$

Mass Relations in Chemistry; Stoichiometry

Worksheet A

A. Circle the true statements and make the false statements true. There can be more than one true statement for each number.

1. For the reaction

$$2\,H_2S\,(g) \; + \; SO_2\,(g) \quad \rightarrow \quad 3\,S\,(s) \; + \; 2\,H_2O\,(g)$$

a. 3 moles of S are produced per mole of H_2S.
b. 1 mole of SO_2 is consumed per mole of H_2S.
c. 1 mole of H_2O is produced per mole of H_2S.
d. The total number of moles of products is always equal to the total number of moles of reactants used.

2. If 1.00 mol of ammonia reacts with 1.00 mol of oxygen according to the reaction

$$4\,NH_3\,(g) \; + \; 5\,O_2\,(g) \quad \rightarrow \quad 4\,NO\,(g) \; + \; 6\,H_2O\,(\ell)$$

a. All the oxygen is consumed.
b. 4.00 mol of NO is produced.
c. 1.50 mol of water is produced.
d. 0.20 mol of ammonia is left over.
e. The statement does not provide enough information to determine percent yield.

3. In the reaction of 2.0 mol CCl_4 with an excess of HF, 1.7 mol CCl_2F_2 is obtained.

$$CCl_4\,(\ell) \; + \; 2\,HF\,(g) \quad \rightarrow \quad CCl_2F_2\,(\ell) \; + \; 2\,HCl\,(g)$$

a. The theoretical yield for CCl_2F_2 is 1.7 mol.
b. The actual yield for CCl_2F_2 is 1.0 mol.
c. The percent yield for the reaction is 85%.
d. Theoretical yield cannot be determined unless the exact amount of HF used is known.

B. Dinitrogen pentaoxide can be produced by the reaction between nitrogen and oxygen.

1. Write a balanced equation for the reaction.

2. Write the same balanced equation pictorially. Use a square to represent **an atom** of nitrogen and a circle to represent **an atom** of oxygen. You may NOT use any letters or numbers, only squares, circles, + and →. *(Remember:* Nitrogen and oxygen are diatomic molecules.)

3. If there are six molecules of both nitrogen and oxygen, show pictorially a before and after representation of the reaction.

4. A reaction uses 4.50 grams of oxygen and an excess of nitrogen. How many grams of dinitrogen pentaoxide are produced?

5. Another reaction uses 3.87 g of both nitrogen and oxygen.
    a. How many grams of dinitrogen pentaoxide are produced?
    b. How many grams of the reactant in excess is left over?
    c. If 3.87 g of dinitrogen pentaoxide is obtained after the experiment, what is the percent yield?

Mass Relations in Chemistry; Stoichiometry

Worksheet B

A. Write a balanced equation to represent the following reactions. Use only smallest whole number coefficients.
   1. When ammonia gas is burned in oxygen, nitrogen oxide gas and water are formed.
   2. When hydrogen sulfide (H_2S) gas reacts with sulfur dioxide gas, solid sulfur (S (s)) and steam are obtained.

B. When 2.50 mol of ammonia, NH_3, reacts with an excess of oxygen, 1.50 mol of NO_2 are obtained. The equation for the reaction is
$$4\,NH_3\,(g)\ +\ 7\,O_2\,(g)\ \rightarrow\ 4\,NO_2\,(g)\ +\ 6\,H_2O\,(\ell)$$
Circle the true statements about the reaction and make the false statements true.
   1. The theoretical yield for NO_2 is 1.50 mol.
   2. The theoretical yield for H_2O is 67.5 g.
   3. The percent yield for the reaction is 60.0%.
   4. The theoretical yield cannot be determined unless the exact amount of oxygen used is given.
   5. It is impossible, given the information above, to calculate how much O_2 is unreacted.
   6. Using the same percent yield data, a student would need 4.00 moles of NH_3 and an excess of oygen to produce 4.00 moles of NO_2.
   7. At the end of the reaction, no NH_3 is theoretically left unreacted.

C. Squares represent atoms A and circles represent atoms B.
   1. Write the reaction that is represented by the picture below.

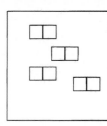 + 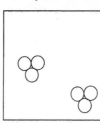 →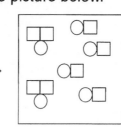

2. Draw a pictorial representation of the same reaction where one starts with 6 molecules of ☐☐ and 5 molecules of ⬭ .

D. When solid silicon tetrachloride reacts with water, solid silicon dioxide and hydrogen chloride gas are formed.

1. Write a balanced equation (using smallest whole number coefficients) for the reaction.

2. In an experiment, 25.0 g of silicon tetrachloride are treated with ten grams of water. How many grams of each reactant and product could you theoretically expect after the reaction is complete?

3. The amount of silicon dioxide recovered from the experiment in (2) is 5.36 g. What is the % yield?

Mass Relations in Chemistry; Stoichiometry

Worksheet C

A. Circle the correct answer(s):

1. When phosphine, PH_3 (g), is burned in oxygen, tetraphosphorus decaoxide and steam are formed. The sum of the coefficents on the product side of the equation is

    a. 5   b. 6   c. 7   d. 8

2. Consider the mixture of iodine gas and chlorine gas represented in the box below. (I atoms are represented as squares, chlorine atoms as circles.)

    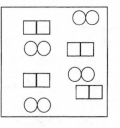

    What will the contents of the box look like if the molecules undergo the reaction

    $$I_2 \text{ (g)} + 3\,Cl_2 \text{ (g)} \rightarrow 2\,ICl_3 \text{ (g)}$$

    a.   b. 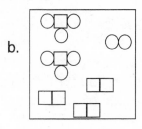   c.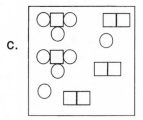

    d.   e.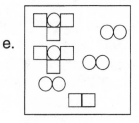

3. A weighed sample of iron is added to liquid bromine and allowed to react completely. The reaction produces a single product which is weighed. The experiment is repeated several times with different masses of iron but with the same mass of bromine. The results are summarized in the following graph.

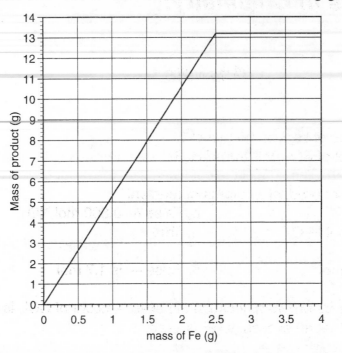

Circle the statements which best describe the experiments.

a. When 1.00 g of iron are added to the bromine, iron is the limiting reactant.

b. When 3.50 g of iron are added to the bromine, there is an excess of bromine.

c. When 2.50 g of iron are added to the bromine, both reactants are used up.

d. When 2.00 g of iron are added to the bromine, 10.0 grams of product are obtained. The % yield for the experiment must therefore be 20.0%.

B. Consider the experiment described in A(3) above.

1. Assume that 100% yield was obtained throughout. How many grams of bromine were used? (*Hint:* Remember the Law of Conservation of Mass.)

2. What is the mole ratio of liquid bromine to iron for the reaction?

3. What is the simplest formula of the product?

C. Methyl benzoate, $C_8H_8O_2$, is prepared by reacting methanol, CH_4O, with benzoic acid, $C_7H_6O_2$.

$$C_7H_6O_2\,(s)\ +\ CH_4O\,(\ell)\ \longrightarrow\ C_8H_8O_2\,(s)\ +\ H_2O\,(\ell)$$

In an experiment, 24.4 g of benzoic acid ($\mathcal{M}$ = 122.1 g/mol) were reacted with 70.0 mL of methanol (d = 0.791 g/mL, $\mathcal{M}$ = 32.04 g/mol).

1. Assuming 100% yield what species are present after reaction is complete?

2. Assuming 100% yield, how much of each reactant is present after reaction is complete?

3. After reaction, 21.6 g of methyl benzoate ($\mathcal{M}$ = 136.1 g/mol) are obtained. What was the percent yield of the experiment?

Mass Relations in Chemistry; Stoichiometry

Answers

Worksheet A

A. 1. a. False — 2 moles of H_2S or 1.5 mol of S
   b. False — 1/2 mol of SO_2 or 2 mol of H_2S
   c. True
   d. False — <u>grams</u> of product ... <u>grams</u> of reactant
   2. a. True   b. False — 0.800 mol NO
   c. False — 1.20 mol H_2O d. True
   e. True
   3. a. False — is 2.0 mol   b. False — is 1.7 mol
   c. True
   d. False — can be determined because the exact amount of CCl_4 is known and it is given that an excess of HF is used.

B. 1. $2 N_2 (g) + 5 O_2 (g) \rightarrow 2 N_2O_5 (g)$

2.

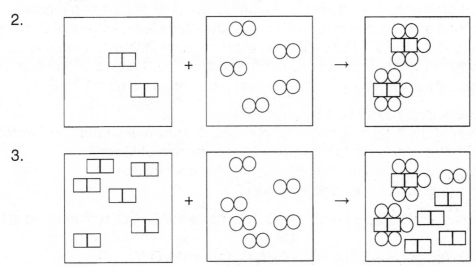

3.

4. 6.08 g N_2O_5
5. a. 5.23 g N_2O_5 b. 2.51 g N_2 left over   c. 74.0%

Worksheet B

A. 1. $4\,NH_3\,(g)\ +\ 5\,O_2\,(g)\ \rightarrow\ 4\,NO\,(g)\ +\ 6\,H_2O\,(\ell)$

2. $2\,H_2S\,(g)\ +\ SO_2\,(g)\ \rightarrow\ 3\,S\,(s)\ +\ 2H_2O\,(g)$

B. 1. False — 2.50 mol   2. True

    3. True   4. False — <u>can</u> be ... <u>because</u> the <u>exact</u> ... of <u>NH₃</u>

    5. True   6. False — 6.67 moles

    7. True

C. 1. $4\,A_2\ +\ 2\,B_3\ \rightarrow\ 2\,A_2B\ +\ 4\,AB$ or $2\,A_2\ +\ B_3\ \rightarrow\ A_2B\ +\ 2\,AB$

    2.

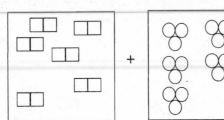 + 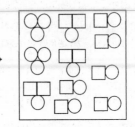 →

D. 1. $SiCl_4\,(s)\ +\ 2\,H_2O\,(\ell)\ \rightarrow\ SiO_2\,(s)\ +\ 4\,HCl\,(g)$

    2. 0.0 g $SiCl_4$,   4.70 g H_2O,   8.84 g SiO_2,   21.5 g HCl

    3. 60.6%

Worksheet C

A. 1. c   2. b   3. a, c

B. 1. 10.7 g Br_2 2. 3 Br_2 : 2 Fe   3. $FeBr_3$

C. 1. $C_8H_8O_2$, H_2O , CH_4O

    2. 0.0 g $C_7H_6O_2$; 61.9 mL CH_4O

    3. 79.4%

Reactions in Aqueous Solutions

Worksheet A

A. Write a balanced net ionic equation using smallest whole number coefficients for the reaction between aqueous solutions of
   1. barium hydroxide and aluminum chloride.

   2. chromium(III) sulfate and strontium sulfide.

   3. ammonia and hydrobromic acid.

   4. perchloric acid and sodium hydroxide.

   5. hydrocyanic acid (HCN) and lithium hydroxide.

B. Consider the following unbalanced equation:

$$WO_3\,(s) + Sn^{2+}\,(aq) \rightarrow W_3O_8\,(s) + Sn^{4+}\,(aq)$$

   1. Write the oxidation number of each element in the reaction.

   2. What is the element that is oxidized? _____

   3. What is the element that is reduced? _____

   4. What species is the reducing agent? _____

   5. What species is the oxidizing agent? _____

   6. Write balanced reduction and oxidation half-reactions in both acid and base media.

   7. Write a balanced net ionic equation (using smallest whole number coefficients) for the reaction in both acid and base media.

C. How many grams of barium hydroxide are required to prepare 250.0 mL of a 0.3340 M $Ba(OH)_2$ solution?

D. How many grams of aluminum sulfide are obtained from 38.00 mL of 0.2000 M aluminum nitrate and an excess of sodium sulfide?

E. Tin(II) ions react with periodate ions in acid medium to give iodide and tin(IV) ions. Twenty-five mL of a 0.250 M solution of tin(II) nitrate react with 45.00 mL of 0.100 M solution of sodium periodate. Assume100% yield and additive volumes.
   1. Write a net ionic equation for the reaction.

   2. What is the concentration of tin(IV) ions after reaction is complete?

   3. What is the concentration of tin(II) ions after reaction is complete?

Reactions in Aqueous Solutions

Worksheet B

A. Write a balanced net ionic equation for the reaction between 0.1 M aqueous solutions of the following compounds. Use smallest whole number coefficients.
1. nickel(II) chloride and barium hydroxide

2. methylamine (CH_3NH_2) and hydroiodic acid

3. sulfuric acid and manganese(II) nitrate

4. nitric acid and lithium hydroxide

5. sulfurous acid and barium hydroxide

B. Consider the following unbalanced equation:

$$Cl^- \, (aq) \, + \, SO_4^{2-} \, (aq) \, \rightarrow \, S_4O_6^{2-} \, (aq) \, + \, ClO_3^- \, (aq)$$

1. Write the oxidation number of each element in the reaction.

2. What element is oxidized? _____

3. What element is reduced? _____

4. What species is the reducing agent? _____

5. What species is the oxidizing agent? _____

6. Write a balanced net ionic equation for the reaction in basic medium.

C. The boxes on the left are pictorial representations of reactants. Fill in the boxes on the right with pictorial representations of the products.

1.

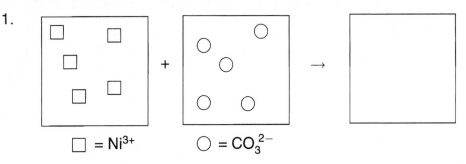

$\square$ = Ni^{3+} $\bigcirc$ = CO_3^{2-}

2.

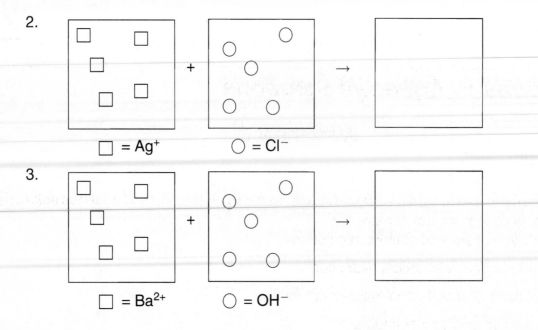

    □ = Ag⁺   ○ = Cl⁻

3.

    □ = Ba²⁺   ○ = OH⁻

D. How many grams of iron(III) hydroxide can be theoretically obtained from the reaction between 25.00 mL of 0.300 M $Sr(OH)_2$ and 25.00 mL of 0.300 M iron(III) nitrate?

E. Five grams of citric acid (M = 180.0 g/mol) are dissolved in 100.0 g of water. The resulting solution reacts with 65.3 mL of strontium hydroxide. The balanced net ionic equation for the reaction is

$$H_3C_5H_5O_7 \text{ (aq)} + 3\,OH^- \text{ (aq)} \longrightarrow C_5H_5O_7^{3-} \text{ (aq)} + 3\,H_2O$$

What is the molarity of the strontium hydroxide solution?

Reactions in Aqueous Solutions

Worksheet C

A. Write a balanced net ionic equation for the reaction between 0.10 M aqueous solutions of the following pairs of compounds.
  1. nickel sulfate and aluminum chloride

  2. nitrous acid and lithium hydroxide

  3. sodium hydroxide and magnesium sulfide

  4. periodic acid and ammonia

  5. hydrochloric acid and sodium fluoride

B. Consider the following unbalanced reaction:

$$S_8\,(s)\ +\ Cr_2O_7^{2-}\,(aq)\ \rightarrow\ SO_4^{2-}\,(aq)\ +\ Cr_2O_3\,(s)$$

  1. What is the oxidizing agent? _____

  2. What species is the reducing agent? _____

  3. Write balanced reduction and oxidation half-reactions in both acid and base media.

  4. Write a balanced net ionic equation (using smallest whole number coefficients) for the reaction in both acid and base media.

C. Write balanced equations to represent the reactions (if they occur) represented pictorially below.

  1.

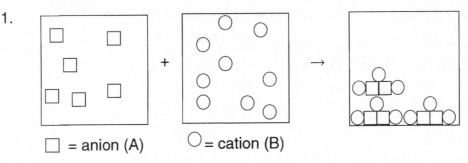

  □ = anion (A)   ○ = cation (B)

2.

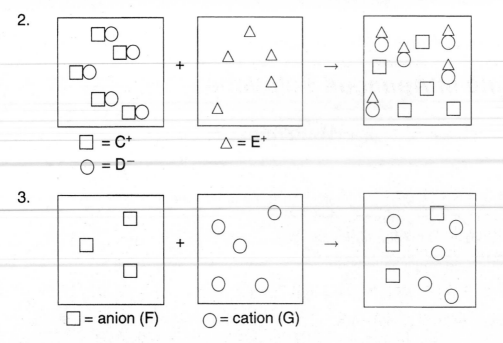

□ = C⁺   △ = E⁺

○ = D⁻

3.

□ = anion (F)   ○ = cation (G)

D. Cyanide ions (CN^-) reduce permanganate ions in basic medium to manganese(IV) oxide. The cyanide ions are oxidized to cyanate ions, CNO^-. The balanced equation for this redox reaction is

$$2\,MnO_4^-\,(aq) + 3\,CN^-\,(aq) + H_2O \rightarrow 3\,CNO^-\,(aq) + 2\,MnO_2\,(s) + 2\,OH^-\,(aq)$$

When a 0.2500 M solution of calcium cyanide is added to an excess of permanganate ions, 12.32 g of MnO_2 ($\mathcal{M}$ = 86.94 g/mol) are obtained. What volume of calcium cyanide is added? Assume 100% yield.

E. When a solution of sodium oxalate ($Na_2C_2O_4$) is added to a solution of lanthanum(III) chloride, lanthanum oxalate ($\mathcal{M}$ = 541.86 g/mol) precipitates. The balanced equation for the reaction is

$$3\,C_2O_4^{2-}\,(aq) + 2\,La^{3+}\,(aq) \rightarrow La_2(C_2O_4)_3\,(s)$$

Thirty mL of 0.2000 M sodium oxalate are added to to 25.00 mL of 0.1500 M $LaCl_3$.

1. How many grams of $La_2(C_2O_4)_3$ (s) are obtained, assuming 100% yield?

2. What are the concentrations (M) of La^{3+}, $C_2O_4^{2-}$, Na^+, and Cl^- after reaction is complete? Assume that volumes are additive.

Reactions in Aqueous Solutions

Answers

Worksheet A

A. 1. Al^{3+} (aq) + $3\,OH^-$ (aq) $\rightarrow$ $Al(OH)_3$ (s)
   2. $2\,Cr^{3+}$ (aq) + $3\,S^{2-}$ (aq) $\rightarrow$ Cr_2S_3 (s)
      Sr^{2+} (aq) + SO_4^{2-} (aq) $\rightarrow$ $SrSO_4$ (s)
   3. NH_3 (aq) + H^+ (aq) $\rightarrow$ $NH_4{}^+$ (aq)
   4. H^+ (aq) + OH^- (aq) $\rightarrow$ H_2O
   5. HCN (aq) + OH^- (aq) $\rightarrow$ CN^- (aq) + H_2O

B. 1. $W = +6$, $O = -2$, $Sn = +2$; $W = +16/3$, $O = -2$, $Sn = +4$
   2. Sn   3. W   4. Sn^{2+} 5. WO_3
   6. <u>acidic medium</u>:
      Oxidation half-reaction: Sn^{2+} (aq) $\rightarrow$ Sn^{4+} (aq) + $2\,e^-$
      Reduction half-reaction: $2\,H^+$ (aq) + $3\,WO_3$ (s) + $2\,e^-$ $\rightarrow$ W_3O_8 (s) + H_2O
      <u>basic medium</u>:
      Oxidation half-reaction: Sn^{2+} (aq) $\rightarrow$ Sn^{4+} (aq) + $2\,e^-$
      Reduction half-reaction: H_2O + $3\,WO_3$ (s) + $2\,e^-$ $\rightarrow$ W_3O_8 (s) + $2\,OH^-$ (aq)
   7. <u>acidic medium</u>: Sn^{2+} (aq) + $2\,H^+$ (aq) + $3\,WO_3$ (s) $\rightarrow$ Sn^{4+} (aq) + W_3O_8 (s) + H_2O
      <u>basic medium</u>: Sn^{2+} (aq) + H_2O + $3\,WO_3$ (s) $\rightarrow$ Sn^{4+} (aq) + W_3O_8 (s) + $2\,OH^-$ (aq)

C. 14.30 g

D. 0.5705 g

E. 1. $4\,Sn^{2+}$ (aq) + $8\,H^+$ (aq) + IO_4^- (aq) $\rightarrow$ $4\,Sn^{4+}$ (aq) + I^- (aq) + $4\,H_2O$
   2. $[Sn^{4+}] = 0.0893$ M   3. $[Sn^{2+}] = 0$ M

Worksheet B

A. 1. Ni^{2+} (aq) + $2\,OH^-$ (aq) $\rightarrow$ $Ni(OH)_2$ (s)
   2. CH_3NH_2 (aq) + H^+ (aq) $\rightarrow$ $CH_3NH_3^+$ (aq)
   3. no reaction
   4. H^+ (aq) + OH^- (aq) $\rightarrow$ H_2O
   5. H_2SO_3 (aq) + OH^- (aq) $\rightarrow$ HSO_3^- (aq) + H_2O

B. 1. $S = +6$, $Cl = -1$, $O = -2$; $S = +2.5$, $Cl = +5$, $O = -2$
   2. Cl   3. S   4. Cl^- 5. SO_4^{2-}
   6. $12\,SO_4^{2-}$ (aq) + $7\,Cl^-$ (aq) + $9\,H_2O$ $\rightarrow$ $3\,S_4O_6^{2-}$ (aq) + $7\,ClO_3^-$ (aq) + $18\,OH^-$ (aq)

C. 1. 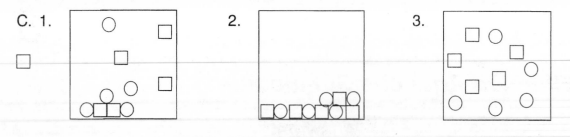 2. 3.

D. 0.534 g

E. 0.638 M

Worksheet C

A. 1. no reaction
   2. HNO_2 (aq) + OH^- (aq) → NO_2^- (aq) + H_2O
   3. Mg^{2+} (aq) + $2\,OH^-$ (aq) → $Mg(OH)_2$ (s)
   4. H^+ (aq) + NH_3 (aq) → NH_4^+ (aq)
   5. H^+ (aq) + F^- (aq) → HF (aq)

B. 1. $Cr_2O_7^{2-}$ 2. S_8
   3. <u>acidic medium</u>:
      Oxidation half-reaction: S_8 (s) + $32\,H_2O$ → $8\,SO_4^{2-}$ (aq) + $48\,e^-$ + $64\,H^+$ (aq)
      Reduction half-reaction: $8\,H^+$ (aq) + $Cr_2O_7^{2-}$ (aq) + $6\,e^-$ → Cr_2O_3 (s) + $4\,H_2O$
      <u>basic medium</u>:
      Oxidation half-reaction: S_8 (s) + $64\,OH^-$ (aq) → $8\,SO_4^{2-}$ (aq) + $48\,e^-$ + $32\,H_2O$
      Reduction half-reaction: $4\,H_2O$ + $Cr_2O_7^{2-}$ (aq) + $6\,e^-$ → Cr_2O_3 (s) + $8\,OH^-$ (aq)
   4. <u>acidic medium</u>: S_8 (s) + $8\,Cr_2O_7^{2-}$ (aq) → $8\,SO_4^{2-}$ (aq) + $8\,Cr_2O_3$ (s)
      <u>basic medium</u>: S_8 (s) + $8\,Cr_2O_7^{2-}$ (aq) → $8\,SO_4^{2-}$ (aq) + $8\,Cr_2O_3$ (s)

C. 1. $3\,B^{2+}$ (aq) + $2\,A^{3-}$ (aq) → B_3A_2 (s)
   2. CD (aq) + E^+ (aq) → ED (aq) + C^+ (aq)
   3. no reaction

D. 0.4251 L

E. 1. 1.016 g
   2. $[La^{3+}] = 0$ M; $[C_2O_4^{2-}] = 0.00682$ M; $[Cl^-] = 0.2045$ M; $[Na^+] = 0.2182$ M

Gases

Worksheet A

A. Draw the following:
1. A balloon filled with 10 molecules of Ne at 25°C and then draw the same balloon in a room where the temperature is 225°C.

2. A sealed steel tank with 10 molecules of H_2 at 25°C and then draw the same sealed steel tank in a room where the temperature is −125°C. (Hydrogen gas condenses to a liquid at −253°C.)

3. The same steel tank at the same two temperatures with the same contents as #2, only this time attach a pressure gauge to the tank.

B. Calculate the molar mass of a gas with density 8.10 g/L at 25°C and 1.00 atm pressure.

C. Ammonia is prepared by reacting nitrogen and hydrogen gases according to the following equation:

$$N_2\,(g)\ +\ 3\,H_2\,(g)\ \longrightarrow\ 2\,NH_3\,(g)$$

If 4.0 L of nitrogen and 4.0 L of hydrogen are combined at constant temperature and pressure, how many liters of ammonia are produced?

D. At 300 K, gas tank A and gas tank B each contain two moles of gas. Tank A contains nitrogen gas, while tank B has oxygen gas.
1. Compare the kinetic energy of the molecules in tanks A and B.

2. Compare the velocity of the molecules in tanks A and B.

3. Compare the time it takes for the molecules to effuse out of tanks A and B.

E. When iron(III) sulfide is roasted in oxygen, iron(III) oxide and sulfur dioxide gas are formed. How many liters of oxygen gas at 25°C and 756 mm Hg are required to roast one kilogram of iron(III) sulfide?

Gases

Worksheet B

A. Circle the true statement(s). If the statement is false, rewrite it to make it true.

1. When 2.0 L of hydrogen gas are combined with 2.0 L of oxygen gas at 25°C and 1.0 atm,

$$2 H_2 (g) + O_2 (g) \rightarrow 2 H_2O (g)$$

4.0 L of steam are obtained.

2. A gas at 25°C in a 10.0 L sealed stainless steel tank has a pressure of 1.00 atm. When the temperature in the tank is doubled to 50°C,
   a. the volume of the tank doubles.
   b. the pressure in the tank doubles.
   c. the number of moles of gas doubles.
   d. the kinetic energy of the molecules increases.
   e. the velocity of the molecules stays the same.

3. A tank has a total pressure of 1.00 atm. It contains 25.0 g of oxygen and 25.0 g of sulfur dioxide. The partial pressure of oxygen in the tank is 0.500 atm.

B. What is the density of CO_2 (g) at 745 mm Hg and 32°C?

C. A gas has a density of 1.10 g/L at 27°C and 0.750 atm. What is its molar mass?

D. Fifteen mL of HCl is added to aluminum metal producing 125 mL of hydrogen gas at 25°C and 1.00 atm. What is the molarity of the HCl added?

E. Consider two bulbs connected by a valve. Bulb A has a volume of 200.0 mL and contains nitrogen gas at a pressure of 0.500 atm. Bulb B has a volume of 1.00 L and contains CO gas at a pressure of 1.00 atm. What is the pressure in the two tanks when the valve is opened? The volume of the connecting tube and valve is negligible. Nitrogen does not react with carbon monoxide.

CHALLENGE PROBLEM: What would the pressure in both tanks be if nitrogen were replaced by oxygen in bulb A under the same conditions?
(*Hint:* $2 CO (g) + O_2 (g) \rightarrow 2 CO_2 (g)$)

Gases

Worksheet C

A. Circle the true statements.

1. According to the kinetic theory of gases
   a. a molecule of H_2 at 100°C and 1 atm has exactly the same velocity as a molecule of oxygen at the same conditions.
   b. the average translational kinetic energy of a mole of oxygen molecules is about 1/4 that of a mole of hydrogen molecules under the same conditions of temperature and pressure.
   c. ten molecules of hydrogen will exert the same pressure as ten molecules of oxygen when volume and temperature are the same.
   d. the observed pressure in gases is due to the collisions of molecules with each other.

2. According to the Ideal Gas Law
   a. density increases with pressure.
   b. molar mass increases with temperature.
   c. for a fixed amount of gas, the quantity PV/T is a constant.
   d. when T and n are kept constant, doubling the volume doubles the pressure.

3. A steel cylinder contains 0.20 mol H_2 ($\mathcal{M} = 2.0$ g/mol) and 0.10 mol He ($\mathcal{M} = 4.0$ g/mol). At 300 K, the cylinder has a total pressure of 6.0 atm.
   a. The partial pressure of hydrogen is 4.0 atm.
   b. The total volume of the cylinder cannot be determined from the data given.
   c. If the contents of the cylinder are transferred without loss or change in temperature to a cylinder with twice the volume of the original cylinder, the mol fraction of each gas will remain the same, but their partial pressures will change.
   d. If 2.00 g of neon gas ($\mathcal{M} = 20.0$ g/mol) replaces all the hydrogen gas, the pressure in the cylinder will increase.

B. Consider a sealed tank kept at a constant temperature. The tank contains 8 moles of neon, 2 moles of methane (CH_4) and 1 mole of oxygen gas.

   1. Which gas in the tank has the highest partial pressure?
   2. Which gas in the tank will effuse out first if a tiny hole is drilled into the tank?
   3. Which gas in the tank has the highest mass % of the mixture?
   4. If 4 moles of neon were replaced by 4 moles of SO_2 gas, would the pressure change and if so in which direction?
   5. If 4 moles of neon were replaced by 8.0 g of SO_2 gas, would the pressure change and if so in which direction?

C. Consider helium.
   1. What is its density at 27°C and 1.25 atm?

   2. What is its average speed at 27°C in miles/hr?

   3. Ten g of helium are pumped into an empty balloon at 27°C and 1.00 atm. What is the volume of the balloon?

D. Acetylene, C_2H_2, can be prepared by combining calcium carbide, CaC_2, and water according to the following reaction.

$$CaC_2 \text{ (s)} + 2\,H_2O \text{ (}\ell\text{)} \rightarrow C_2H_2 \text{ (g)} + Ca(OH)_2 \text{ (s)}$$

To produce the acetylene, the procedure suggests that 50.0% more water than required be added to the calcium carbide. What mass of calcium carbide and volume of water (d = 1.00 g/mL) must be combined according to the procedure, so that 10.0 L of acetylene at 27°C and 1.00 atm are produced?

Gases

Answers

Worksheet A

A. 1. at 25°C: at 225°C:

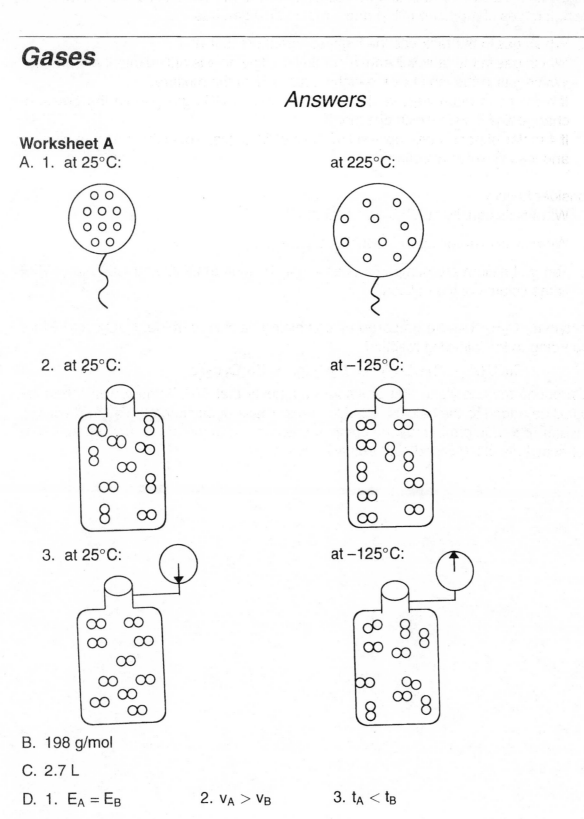

2. at 25°C: at −125°C:

3. at 25°C: at −125°C:

B. 198 g/mol

C. 2.7 L

D. 1. $E_A = E_B$ 2. $v_A > v_B$ 3. $t_A < t_B$

E. 532 L

Worksheet B
A. 1. False — <u>2.0 L</u> of steam
    2. a. F — ... tank <u>remains the same</u>
       b. F — ... tank <u>increases but does not double</u>
       c. F — ... of gas <u>remains the same</u>
       d. True
       e. F — ... molecules <u>increases</u>
    3. False — <u>0.666</u> atm

B. 1.72 g/L

C. 36.1 g/mol

D. 0.681 M

E. 0.917 atm   Challenge Problem: 0.833 atm

Worksheet C
A. 1. c   2. a, c   3. a, c

B. 1. Ne   2. CH_4 3. Ne
   4. no   5. Yes; pressure would decrease

C. 1. 0.203 g/L   2. 3.06×10^3 mi/h   3. 61.6 L

D. 26.0 g CaC_2 ;   21.9 mL H_2O

Electronic Structure and the Periodic Table

Worksheet A

A. Circle the true statements.

1. The orbital pictured below belongs to an orbital of the subshell

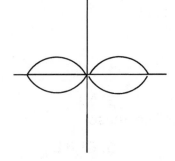

    a. $\ell = 0$ b. $\ell = 1$ c. $\ell = 2$ d. none of these

2. For the quantum number n = 3
   a. the quantum number m_ℓ can only be 0.
   b. the quantum number ℓ cannot be larger than 2.
   c. the quantum number m_s must be 1/2.
   d. there can be five possible values for m_ℓ.

3. The m_ℓ quantum number for an electron in a 5d orbital
   a. can have any value less than 5.
   b. may be 0.
   c. may be –3.
   d. may be + 1/2 or – 1/2.
   e. has to be 5.

4. The number of 2p electrons in an atom of Cl is
       a. 0   b. 2   c. 5   d. 6

5. An atom of As has
       a. 5 e⁻ in the 4p subshell.   b. 10 e⁻ in the 4d subshell.
       c. 6 e⁻ in the 3p subshell.   d. 3 e⁻ in the 4p subshell.

B. Answer the following questions:
1. How many unpaired electrons are in an atom of scandium?

2. What is the electron configuration and orbital diagram for Ni?

3. What is the abbreviated electron configuration for Ni^{2+}? Ni^{3+}?

4. What is the electronic configuration for N^{3-}?

5. Write the symbol for the halide (halogen ion) that is isoelectronic to xenon.

C. Give the symbol of the element
1. in Group 14 that has the smallest atomic radius.

2. in Period 3 that has the highest ionization energy.

3. in Group 16 that has the lowest electronegativity.

D. What frequency (in s^{-1}) is associated with the transition from n = 4 to n = 2?

E. On another planet, the rules for quantum numbers are as follows:

$$n = 1, 2, 3, 4, ..., + \infty$$
$$\ell = 2, 3, 4, ..., n + 1 \qquad \text{where} \qquad \ell = 2 \text{ is designated as } a$$
$$\ell = 3 \text{ is designated as } b$$
$$\ell = 4 \text{ is designated as } c$$
$$\ell = 5 \text{ is designated as } d$$
$$m_\ell = 0, 1, 2, ..., \ell$$
$$m_s = +\tfrac{1}{2}, 0, -\tfrac{1}{2}$$

(All other rules are the same as those on planet Earth.)
1. How many electrons can go into n = 1?, n = 2?

2. How many electrons are there in each orbital (m_ℓ)?

3. Write the electron configuration for Cl and Br according to the rules of the other planet.

$\lambda = c/\nu$

$c = 2.998 \times 10^8$ m/s

$E = -R_H/n^2$

$h = 6.628 \times 10^{-34}$ J-s

$E = h\nu$

$R_H = 2.180 \times 10^{-18}$ J

Electronic Structure and the Periodic Table

Worksheet B

A. Circle the true statements.

1. When an electron in a hydrogen atom changes its principal quantum number from 1 to 3
   a. it absorbs energy, usually in the form of light.
   b. it gives off energy.
   c. $\Delta E > 0$.
   d. its electron configuration at that excited state could be $3s^1$.

2. A 3p subshell
   a. can have an electron with a quantum number of $\ell = 2$.
   b. can hold a maximum of 6 electrons.
   c. can have an electron with a quantum number $m_\ell = -2$.
   d. can have a maximum of 3 unpaired electrons for an atom in the ground state.

3. An atom has electrons in both orbitals A and B.

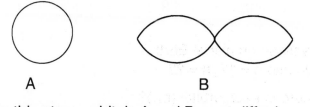

   A   B

   In this atom, orbitals A and B must differ in
   a. n   b. ℓ c. m_ℓ d. m_s

4. An important factor in determining the size of the halogens in their normal ground state is
   a. the number of unpaired electrons.
   b. the number of electrons in the outermost shell.
   c. the number of principal energy levels.
   d. the effective nuclear charge (Z_{eff}).
   e. the number of electrons in $\ell = 1$.

5. The first two quantum numbers (n, ℓ) of the highest energy electron of scandium in the ground state are
   a. 3, 0   b. 3, 1   c. 3, 2   d. 4, 0   e. 4, 1

B. Answer the following questions:
1. How many electrons with quantum number $\ell = 0$ are there in the ground state electron configuration for rubidium?

2. How many electrons can have both quantum numbers $n = 3$ and $\ell = 2$?

3. How many unpaired electrons are there for bromine in its unexcited state?

4. What is the abbreviated electronic configuration for V^{3+}?

5. What element is represented by the electron configuration $1s^2\, 2s^1\, 2p^5\, 3s^2\, 3p^2$? Is this an electron configuration for this element in its unexcited state?

C. Give the symbol of the element
1. in Group 16 that is the most electronegative.

2. in Period 3 that has the largest radius.

3. in Group 1 (not counting hydrogen) that has the highest ionization energy

4. that is the most electronegative.

D. An electron is at $n = 4$. To what n will it drop if it gives off light with a wavelength of 486.1 nm?

E. In another planet, three electrons are allowed in each orbital ($m_s = +\frac{1}{2}, 0, -\frac{1}{2}$). If their physical laws are similar to ours, construct Period 2 of our periodic table consistent with their quantum laws.

$\lambda = c/\nu$

$c = 2.998 \times 10^8$ m/s

$E = -R_H/n^2$

$h = 6.628 \times 10^{-34}$ J-s

$E = h\nu$

$R_H = 2.180 \times 10^{-18}$ J

Electronic Structure and the Periodic Table

Worksheet C

A. Circle the correct choice(s).

1. An acceptable designation for an atomic orbital is
   a. 4g   b. 5g   c. 3f   d. 6d   e. 4f

2. This element has unpaired electrons in its ground state electron configuration.
   a. Ca   b. V   c. Co   d. Cu   e. Kr

3. This element is isoelectronic with krypton.
   a. Y^{3+} b. Br^- c. S^{2-} d. Ti^{2+} e. Sc^{3+}

4. This is an allowed state for an electron in an atom.
   a. $n = 3, \ell = 2, m_\ell = -2$ b. $n = 3, \ell = 1, m_\ell = 0$
   c. $n = 3, \ell = 0, m_\ell = -1$ d. $n = 3, \ell = 2, m_\ell = 0$
   e. $n = 3, \ell = 3, m_\ell = -2$

5. An s orbital
   a. is pie shaped about the nucleus.
   b. forms concentric circles from the nucleus.
   c. is spherical about the nucleus.
   d. has an uncertain shape.
   e. has quantum number $\ell = 0$.
   f. is none of the above.

6. Which of these statements are true?
   a. Light has wave properties.
   b. Infrared radiation corresponds to higher energy radiation than visible light.
   c. Atoms emit light only at discrete, well-defined frequencies.
   d. The Pauli exclusion principle requires that only two electrons occupy a given orbital.

B. Consider an iron, Fe, atom.
1. Write its unexcited ground state electron configuration.

2. Write its orbital diagram.

3. Write the abbreviated ground state configuration for Fe^{2+}.

4. Compare the size of Fe with Fe^{2+}; Fe^{2+} with Fe^{3+}.

5. What is the value of the highest n in the ground state?

6. What is the value of the highest ℓ?

C. Give the symbol of the element
1. of lowest atomic number which could have the following excited state:
$1s^2\ 2s^2\ 2p^6\ 3d^1$

2. in Group **14** which has the largest atomic radius.

3. in Group **16** that has the highest electronegativity value.

4. in Period 5 which has the lowest ionization energy.

D. What are the energy (in J/photon and kJ/mol), frequency, and wavelength associated with an n = 5 to n = 1 transition?

E. On another planet, the rules for quantum numbers are as follows:

$$n = 1, 2, 3, 4, ..., +\infty$$
$$\ell = 1, 2, 3, 4 ..., n$$

where $\ell = 1$ is designated as h
$\ell = 2$ is designated as i
$\ell = 3$ is designated as j
$\ell = 4$ is designated as k

$$m_\ell = 0, 1, 2, ..., \ell - 1$$
$$m_s = +\tfrac{1}{2}$$

(All other rules are the same as those on planet Earth.)
1. How many electrons can go into n = 1?, n = 3?

2. Write the electron configuration for P according to the rules of the other planet.

$\lambda = c/\nu$ $E = -R_H/n^2$ $E = h\nu$
$c = 2.998 \times 10^8$ m/s $h = 6.628 \times 10^{-34}$ J-s $R_H = 2.180 \times 10^{-18}$ J

Electronic Structure and the Periodic Table

Answers

Worksheet A

A 1. b   2. b, d   3. b   4. d   5. c, d

B. 1. one

2. $1s^2$ $2s^2$ $2p^6$ $3s^2$ $3p^6$ $4s^2$ $3d^8$

      (↑↓)   (↑↓)   (↑↓)(↑↓)(↑↓)   (↑↓)   (↑↓)(↑↓)(↑↓)   (↑↓)   (↑↓)(↑↓)(↑↓)(↑)(↑)

3. [Ar] $3d^8$; [Ar] $3d^7$

4. $1s^2\,2s^2\,2p^6$

5. I^-

C. 1. C   2. Ar   3. Po

D. $6.169 \times 10^{14}\ s^{-1}$

E. 1. 9; 21   2. 3   3. Cl: $1a^9\,2a^8$ Br: $1a^9\,2a^9\,2b^{12}\,3a^5$

Worksheet B

A. 1. a, c, d   2. b, d   3. b   4. c   5. c

B 1. 9   2. 10   3. 1   4. $[Ar]3d^2$ 5. Mg; no

C. 1. O   2. Na   3. Li   4. F

D. n = 2

E.

Be	B	C

N	O	F	Ne	Na	Mg	Al	Si	P

Worksheet C

A. 1. b, d, e   2. b, c, d   3. a, b
   4. a, b, d   5. c, e   6. a, c, d

B. 1. $1s^2\ 2s^2\ 2p^6\ 3s^2\ 3p^6\ 4s^2\ 3d^6$

   2. $(\uparrow\downarrow)$ $(\uparrow\downarrow)$ $(\uparrow\downarrow)(\uparrow\downarrow)(\uparrow\downarrow)$ $(\uparrow\downarrow)$ $(\uparrow\downarrow)(\uparrow\downarrow)(\uparrow\downarrow)$ $(\uparrow\downarrow)$ $(\uparrow\downarrow)(\uparrow\ )(\uparrow\ )(\uparrow\ )(\uparrow\ )$

         1s   2s   2p   3s   3p   4s   3d

   3. $[Ar]\ 3d^6$ 4. $Fe > Fe^{2+}$; $Fe^{2+} > Fe^{3+}$

   5. 4   6. 2

C. 1. Na   2. Pb   3. O   4. Rb

D. $E = 2.093 \times 10^{-18}$ J/photon $= 1.260 \times 10^3$ kJ/mol
   $\nu = 3.159 \times 10^{15}$ s^{-1}
   $\lambda = 94.94$ nm

E. 1. 1, 6
   2. $1h^1\ 2h^1\ 2i^2\ 3h^1\ 3i^2\ 3j^3\ 4h^1\ 4i^2\ 4j^2$

Covalent Bonding

Worksheet A

A. Print the matching letters on the blanks provided. There is only one correct answer for each blank. Each structure can be used more than once.

_____ 1. has an atom with three pairs of electrons around it.

_____ 2. has a C atom with sp^2 hybridization.

_____ 3. has a bent shape with an angle of 120°.

_____ 4. has a bent shape with an angle of 109°.

_____ 5. is tetrahedral.

_____ 6. has 2π and 2σ bonds.

_____ 7. is nonpolar.

_____ 8. has a C atom with sp hybridization.

_____ 9. has sp^3d^2 hybridization on the central atom.

_____ 10. has a flat structure with 120° angle and no unshared e^- pairs.

_____ 11. is T-shaped.

_____ 12. is linear.

_____ 13. has a carbon atom with sp^3 hybridization.

_____ 14. is square planar.

_____ 15. can have a resonance structure with a triple bond.

A. CH_3Br B. SeF_2 C. IF_3 D. XeF_3Cl

E. CNO^- F. BF_3 G. SO_2 H. $H_2C=CHCl$

B. Draw Lewis structures for the following species:
    1. CO_3^{2-}

    2. PCl_4^-

    3. SF_4^{2-}

    4. CO_2

    Determine for these species:
    a. resonance structures (if any)
    b. bond angle between the central atom and two bonds in the molecule
    c. geometry
    d. hybridization of the central atom
    e. formal charge of the central atom
    f. number of sigma and pi bonds

C. What is the polarity of the species (A – H) in Part A?

Covalent Bonding

Worksheet B

A. Circle the statement(s) that best answer(s) the question.
1. Which best describes a covalent bond?
   a. An ion electrostatically held to a cation.
   b. Two nuclei sharing a pair of electrons that is located exactly between the nuclei.
   c. Three atoms held by a pair of electrons.
   d. Two atoms sharing an electron pair.

2. What type of bond is formed between the carbon and nitrogen atoms in the HCN molecule?
   a. single covalent   b. double covalent
   c. triple covalent   d. ionic

3. Which species violates the octet rule?
   a. CO   b. NH_3 c. IF   d. BF_3

4. Which of the following bonds would be the most polar?
   a. N–F   b. C–F   c. F–O   d. F–F

B. Consider the molecule 3-aminopropynoic acid, whose Lewis structure is given below:

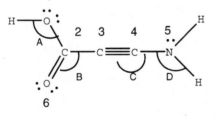

1. Write the hybridization of each of the numbered atoms.

   1: _____   2: _____   3: _____   5: _____

2. Write the bond angle formed between the indicated atoms:

   A: _____   B: _____   C: _____   D: _____

3. Write the number of sigma and pi bonds between the atoms listed below:

   Atoms 1 and 2: _____   _____

   Atoms 2 and 6: _____   _____

   Atoms 3 and 4: _____   _____

C. For the following species

    1. SO_3 2. $SeCl_4$ 3. $XeBr_4$ 4. IBr_2Cl

write
a. the total number of valence electrons.
b. the Lewis structure.
c. the bond angle.
d. the geometry.
e. the polarity.
f. the hybridization of the central atom.
g. the number of sigma and pi bonds.

D. Two possible skeletal strucutres for formaldehyde, CH_2O, are

    H – C – O – H   and   H – C – O
                                           |
                                         H

1. Draw Lewis structures for both skeletons.

2. Based only on formal charge considerations, which is the more likely structure?

Covalent Bonding

Worksheet C

A. State whether the following statements are true or false. Rewrite the false statements to make them true.
  1. Covalent bonds between atoms consist of paired electrons.
  2. Electronegativity increases when going down a group in the periodic table.
  3. When considering the following bonds for polarity, C–I is the most polar bond.
     C – F, C – Cl, C – Br, C – I
  4. SO_2 is a dipole,. CO_2 is not.. We can therefore say that SO_2 is linear and CO_2 is not.
  5. The triiodide ion, I_3^-, is bent with a bond angle somewhat less than the ideal tetrahedral angle of $109.54°$.
  6. The bond(s) in O=O represent(s) a σ bond and a π bond.
  7. In the molecule SF_4, sulfur has sp^3 hybridization.
  8. Ethylene, $CH_2 = CH_2$, has a π bond.
  9. $NaNO_3$ contains ions and shared electron pairs.
  10. There are always only four pairs of electrons (shared and unshared) around any atom in a molecule with covalent bonds.

B. For the following molecules,

  1. SiO_2 2. SO_2 3. XeF_4 4. SF_4 5. IBr_3

  write their
  a. Lewis structures and resonance structures (if any).
  b. bond angles(s).
  c. polarity.
  d. hybridization of the central atom.
  e. geometry
  f. the number of sigma and pi bonds.

C. Much of the chemistry of living systems revolves around the element phosphorus. The structure below is a Lewis structure for a compound similar to the molecule that holds the nucleic acids (ribonucleic acid, RNA, and deoxyribonucleic acid, DNA) together. Before answering any of the following questions, fill in the structure with any missing unshared electron pairs. There are no pi bonds.

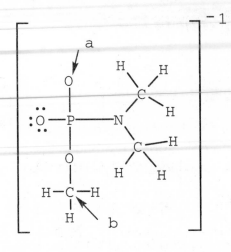

1. What is the geometry around the nitrogen atom?
2. What is the hybridization of the oxygen atom between phosphorus and carbon?
3. What is the formal charge on the oxygen atom marked with a small a?
4. What is the geometry around the phosphorus atom?
5. Which atom(s) has(have) expanded octets?
6. What is the hybridization of all the carbon atoms?
7. How many valence electrons does the nitrogen atom have?
8. What is the approximate value of the bond angle marked with a small b?
9. How many nonpolar bonds are there?
10. How many unshared electron pairs are there?

Covalent Bonding

Answers

Worksheet A

A. 1. F   2. H   3. G   4. B   5. A
   6. E   7. F   8. E   9. D   10. F
  11. C   12. E   13. A   14. D   15. E

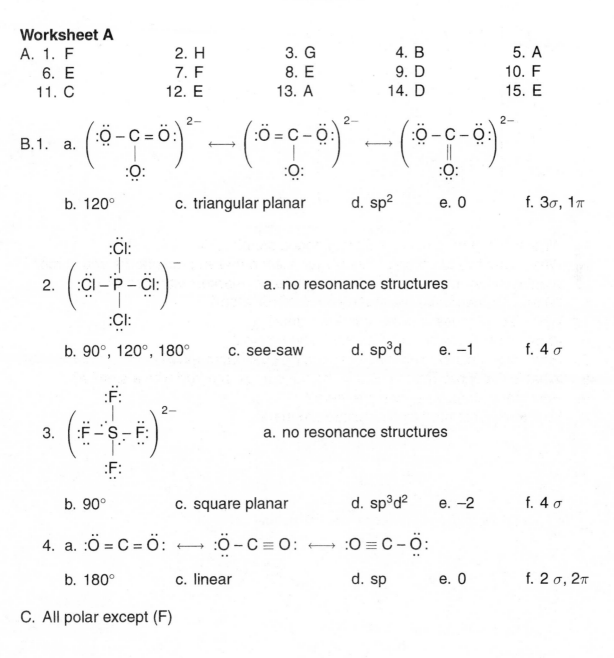

B.1. a.

      b. 120°   c. triangular planar   d. sp^2 e. 0   f. $3\sigma, 1\pi$

2.   a. no resonance structures

      b. 90°, 120°, 180°   c. see-saw   d. sp^3d e. −1   f. 4 σ

3.   a. no resonance structures

      b. 90°   c. square planar   d. sp^3d^2 e. −2   f. 4 σ

4. a. :Ö = C = Ö: ⟷ :Ö − C ≡ O: ⟷ :O ≡ C − Ö:

      b. 180°   c. linear   d. sp   e. 0   f. 2 $\sigma, 2\pi$

C. All polar except (F)

Worksheet B

A. 1. d   2. c   3. d   4. b

B. 1. sp^3 ; sp^2 ; sp ; sp^3 2. 109° ; 120° ; 180° ; 109°
   3. 1 σ; 1 σ, 1 π ; 1 σ, 2 π

C. 1. a. 24   b. :Ö – S = Ö: c. 120°   d. triangular planar
                              |
                             :O:

   e. nonpolar   f. sp^2 g. 3 σ , 1 π

                             :Cl:
                              |
   2. a. 34   b. :Cl – Se – Cl: c. 180°, 90°, 120°   d. see-saw
                              |
                             :Cl:

   e. polar   f. sp^3d g. 4 σ

                             :Br:
                            . | .
   3. a. 36   b. :Br – Xe – Br: c. 180°, 90°   d. square planar
                              |
                             :Br:

   e. nonpolar   f. sp^3d^2 g. 4 σ

   4. a. 28   b. :Cl – I – Br: c. 180°, 90°   d. T–shaped
                              |
                             :Br:

   e. polar   f. sp^3d g. 3 σ

D. 1. H – C̈ = Ö – H   and   H – C = Ö:
                                                   |
                                                   H

   2. C_f C = –1 C_f C = 0
      C_f O = +1 C_f O = 0
   The second structure is the more likely structure.

Worksheet C

A. 1. True   2. False — decreases
    3. False — C – F   4. False — CO_2 is linear, SO_2 is not.
    5. False — linear, 180°   6. True
    7. False — sp^3d 8. True
    9. True   10. False — Most often there are four pairs.

B. 1. a. $:\ddot{O} = Si = \ddot{O}: \longleftrightarrow :\ddot{O} - Si \equiv O: \longleftrightarrow :O \equiv Si - \ddot{O}:$

        b. 180°   c. nonpolar   d. sp   e. linear   f. $2\sigma, 2\pi$

    2. a. $:\ddot{O} - \ddot{S} = \ddot{O}: \longleftrightarrow :\ddot{O} = \ddot{S} - \ddot{O}:$

        b. 120°   c. polar   d. sp^2 e. bent   f. $2\sigma, 1\pi$

    3. a.

$$
\begin{array}{c}
:\ddot{F}: \\
| \\
:\ddot{F} - \overset{\displaystyle \cdot}{\underset{|}{Xe}} - \ddot{F}: \\
| \\
:\ddot{F}:
\end{array}
$$

        no resonance structures

        b. 180°, 90°   c. nonpolar   d. sp^3d^2 e. square planar   f. 4σ

    4. a.

$$
\begin{array}{c}
:\ddot{F}: \\
| \\
:\ddot{F} - \overset{\cdot}{\underset{|}{S}} - \ddot{F}: \\
:\ddot{F}:
\end{array}
$$

        no resonance structures

        b. 120°, 180°, 90°   c. polar   d. sp^3d e. see-saw   f. 4σ

    5. a.

$$
\begin{array}{c}
:\ddot{Br} - \overset{\cdot\cdot}{\underset{|}{I}} - \ddot{Br}: \\
:\ddot{Br}:
\end{array}
$$

        no resonance structures

        b. 180°, 90°   c. polar   d. sp^3d e. T-shaped   f. 3σ

C. 1. triangular pyramid   2. sp^3 3. – 1   4. tetrahedral
    5. none   6. sp^3 7. 5   8. 109.5°
    9. none   10. 9

Thermochemistry

Worksheet A

A. Circle the correct answer(s).

1. An endothermic reaction (in which a solute is dissolved in water) is carried out in a coffee cup calorimeter. Which of the following is(are) NOT true for this process? (There can be more than one answer.)
   a. The temperature of the water increases.
   b. Heat is absorbed by the water.
   c. The enthalpy of the products is higher than the enthalpy of the reactants.
   d. $q_{H_2O} = q_{reaction}$
   e. $q_{reaction} > 0$
   f. $q_{reaction} + q_{H_2O} = 0$

2. The heat of fusion for water (ΔH for ice melting) is +6.00 kJ. Which of the following statements is(are) true for the following reaction?

   $$H_2O\ (\ell) \quad \rightarrow \quad H_2O\ (s)$$

   a. The reaction is endothermic.
   b. The heat flow, q, is > 0.
   c. The enthalpy for $H_2O\ (\ell)$ is higher than that for $H_2O\ (s)$.
   d. Heat is given off by the system.
   e. The temperature of the surroundings decreases.

3. Which statement(s) is(are) true about ΔH?
   a. When $\Delta H > 0$, it means that an endothermic reaction has occured.
   b. ΔH is the amount of heat required to raise the temperature of 1 g of a substance by 1°C.
   c. ΔH for the melting of a pure substance (without change of temperature) is equal in magnitude but opposite in sign for the freezing of that same substance (without change of temperature).
   d. ΔH is equal to ΔH_f° for a reaction in which one mole of a compound is formed from elements in their native states at 1 atm pressure and at a specified temperature.
   e. The unit for ΔH is kJ/mol.

B. Calculate ΔH for the reaction

$$Fe_2CO_3\,(s) + CO\,(g) \rightarrow 2\,FeO\,(s) + CO_2\,(g)$$

Given that

$$Fe_2CO_3\,(s) + 3\,CO\,(g) \rightarrow 2\,Fe\,(s) + 3\,CO_2\,(g) \qquad \Delta H = -26.8\,kJ$$
$$Fe\,(s) + CO_2\,(g) \rightarrow FeO\,(s) + CO\,(g) \qquad \Delta H = 16.5\,kJ$$

C. When 2.50 g of glucose ($C_6H_{12}O_6$) burns in air, carbon dioxide gas and liquid water are formed. 38.9 kJ of heat are also liberated.
   1. What is the heat flow of the reaction if 85.0 g of glucose are burned?

   2. The reaction takes place in a bomb calorimeter where 4.50 g of glucose are burned. What is the heat capacity of the bomb if the temperature of the water in the bomb increases from 22.75°C to 25.32°C?

   3. Using the heat capacity of the bomb calculated in (2), what would the temperature change of the water in the bomb be if 12.0 g of glucose were burned?

   4. How many grams of glucose must be burned to liberate ten kJ of heat?

   5. Write a thermochemical equation for the reaction.

Thermochemistry

Worksheet B

A. Circle the true statements.
1. An exothermic reaction is carried out in a test tube.
   a. Heat is absorbed by the system.
   b. The test tube feels warmer after the reaction is complete.
   c. The enthalpy of the products is higher than the enthalpy of the reactants.
   d. $q_{reaction} > 0$
   e. $q_{reaction} + q_{surroundings} = 0$

2. When the enthalpy of the reactants is greater than the enthalpy of the products
   a. the temperature of the surroundings increases.
   b. $q_{rxn} > 0$ at constant pressure.
   c. the reaction is endothermic.
   d. heat is absorbed by the reaction.
   e. a change in the mass of the reactants can change the value of ΔH but not its sign.

B. Answer the following questions.
1. The heat of vaporization for water (ΔH for water turning to steam at 100°C) is 40.7 kJ. For the following reaction:

$$H_2O \text{ (g)} \quad \rightarrow \quad H_2O \text{ (}\ell\text{)}$$

   a. H_2O (condenses or vaporizes).
   b. q for the reaction is _____ .
   c. ΔH_f° for H_2O (ℓ) ($>, =, <$) ΔH_f° for H_2O (g).
   d. For the above reaction heat is (given off or absorbed) by the system.
   e. The temperature of the surroundings (increases or decreases).

2. Equal masses of liquid A, initially at 100°C, and liquid B, initially at 50°C, are combined in an insulated container. All the heat flow occurs between the two liquids, but the two liquids do not react with each other. The specific heat of liquid A is larger than the specific heat of liquid B. Will the final temperature of the mixture be greater than, equal to, or less than 75°C? (Assume no heat loss.)

C. When 5.00 g of acetone (C_3H_6O) burns in air, carbon dioxide gas and liquid water are formed. Enough heat is liberated to increase the temperature of 1.000 kg of water from 25.0°C to 61.8°C. The specific heat of water is 4.18 J/g-°C

   1.   How many kJ of heat are liberated by the combustion described?

   2.   How many grams of acetone must be burned to liberate 5.00 kJ?

   3.   Write the thermochemical equation for the combustion of acetone.

   4.   What is ΔH_f° for acetone?

   5.   What volume of oxygen at 25°C and 747 mm Hg is consumed when 25.00 kJ of heat are liberated?

Thermochemistry

Worksheet C

A. Which of the following statements are true? Rewrite the false statements to make them true.

1. A positive sign on an enthalpy change means that an endothermic process has occurred.
2. The calorimeter constant is also known as the heat capacity of the calorimeter.
3. The enthalpy of vaporization (ΔH_{vap}) is numerically equal but opposite in sign to the enthalpy of fusion (ΔH_{fus}).
4. The standard state of any substance is the physical state in which it is most stable at 1 atm pressure and 25°C.
5. $q = \Delta H$ at constant volume.
6. $q_{reaction} = q_{surroundings}$
7. $\Delta H_{reactants} - \Delta H_{products} = \Delta H$
8. Bond energy is always a positive quantity.
9. The heat of formation ($\Delta H°$) of water is equal to $\Delta H°$ for the following reaction:

$$2\,H_2\,(g)\, +\, O_2\,(g)\, \rightarrow\, 2\,H_2O\,(\ell)$$

10. Two different metals of equal mass are both heated to 100°C. Each metal is placed in a coffee cup calorimeter containing 25.00 g of H_2O at 22.0°C. The final temperature of the water in the coffee cup with the metal that has the higer specific heat will be higher.

B. A sample of ground beef weighing 5.000 g was burned in a bomb calorimeter giving a temperature rise of 5.71°C. If the calorimeter constant is 1.10×10^4 J/°C, how many nutritional calories does a quarter pound of hamburger made from this ground beef have? (1 Ncal = 4.18 kJ; 1 lb = 454 g)

C. Derive an expression for $\Delta H°$ for the reaction

$$2\,XO\,(s)\, +\, CO_2\,(g)\, \rightarrow\, X_2O_3\,(s)\, +\, CO\,(g)$$

from

$$X\,(s)\, +\, CO_2\,(g)\, \rightarrow\, XO\,(s)\, +\, CO\,(g) \qquad \Delta H_1°$$
$$CO\,(g)\, +\, \tfrac{1}{2}O_2\,(g)\, \rightarrow\, CO_2\,(g) \qquad \Delta H_2°$$
$$2\,X\,(s)\, +\, \tfrac{3}{2}O_2\,(g)\, \rightarrow\, X_2O_3\,(s) \qquad \Delta H_3°$$

D. When one mole of pentaborane, B_5H_9 (g), ignites spontaneously in air, a green flash, solid diboron trioxide, liquid water, and 4507.6 kJ are produced.

   1. Write the thermochemical equation for this reaction.

   2. Given that ΔH_f° for pentaborane is 62.76 kJ/mol, and ΔH_f° for liquid water is − 285.8 kJ/mol, find ΔH_f° for B_2O_3.

   3. Assuming 100% yield, how much heat is given off when 5.00 L of air (21% O_2 by volume) at 756 mm Hg and 25.0°C reacts with an excess of pentaborane?

Thermochemistry

Answers

Worksheet A

A. 1. a, b, d   2. c, d   3. a, c

B. 6.2 kJ

C. 1. -1.32×10^3 kJ   2. 27.2 kJ/°C
   3. 6.86°C   4. 0.643 g
   5. $C_6H_{12}O_6 (s) + 6 O_2 (g) \rightarrow 6 CO_2 (g) + 6 H_2O (\ell)$ $\Delta H = -2.80 \times 10^3$ kJ

Worksheet B

A. 1. b, e   2. a, e

B. 1. a. condenses   b. -40.7 kJ   c. $<$ d. given off   e. increases
   2. greater than

C. 1. 154 kJ
   2. 0.162 g
   3. $C_3H_6O (\ell) + 4 O_2 (g) \rightarrow 3 CO_2 (g) + 3 H_2O (\ell)$ $\Delta H = -1.79 \times 10^3$ kJ
   4. -248 kJ/mol
   5. 1.39 L

Worksheet C

A. 1. T   2. T
   3. F — enthalpy of condensation   4. T
   5. F — constant pressure   6. F — $q_{rxn} = -q_{surr}$
   7. F — $\Delta H = \Delta H_{product} - \Delta H_{reactant}$ 8. T
   9. F — $H_2 (g) + \frac{1}{2} O_2 (g) \rightarrow H_2O (\ell)$ 10. T

B. 341 Ncal

C. $\Delta H = \Delta H_3^o - 2 \Delta H_1^o - 3 \Delta H_2^o$

D. 1. $B_5H_9 (g) + 6 O_2 (g) \rightarrow \frac{5}{2} B_2O_3 (s) + \frac{9}{2} H_2O (\ell)$ $\Delta H = -4507.6$ kJ
   2. -1263.5 kJ/mol
   3. 32 kJ

Liquids and Solids

Worksheet A

A. Circle the correct answer(s).
1. The equilibrium pressure of a pure liquid
   a. remains constant with increasing T.
   b. decreases to half its original value if the volume of the gas phase is doubled, the temperature is kept constant, and liquid is present in the flask.
   c. increases to twice its original value if the volume of the gas phase is halved, the temperature is kept constant, and liquid is present in the flask.
   d. is independent of the volume of the gas phase as long as liquid is present.
   e. decreases to half its original value if the surface area of the liquid is reduced by one-half.
   f. decreases with decreasing T.

2. Bromine, Br_2, boils at 58.8°C, while iodine monochloride, ICl, boils at 97.4°C. The principal reason why ICl boils almost 40°C higher than Br_2 is that
   a. the molar mass of ICl is 162.4 while that of Br_2 is 159.8.
   b. ICl is an ionic compound while Br_2 is molecular.
   c. Dispersion forces for ICl are much higher than those for Br_2.
   d. ICl is polar while Br_2 is nonpolar.
   e. I and Cl are more electronegative than Br.

3. An ionic solid
   a. does not conduct electricity.
   b. is always soluble in water.
   c. can conduct electricity when the solid is melted.
   d. can sometimes have some covalent bonds in its structure.
   e. has an equal number of cations and anions.

B. A certain compound has a normal melting point of 41°C and a normal boiling point of 123°C. The triple point of the compound is 39°C and 85 mm Hg. The vapor pressure of the compound at 50°C is 120 mm Hg.
1. Draw the phase diagram for the compound and label the phases, the triple point, the vapor pressure curve, the melting point curve, and the sublimation curve.
2. Is the solid phase denser than the liquid phase?
3. What happens if P is kept at 500 mm Hg and T is increased from 50°C to 150°C?
4. What happens if T is kept at 20°C and P is increased from 20 mm Hg to 150 mm Hg?
5. If 4.60 g of the compound (M = 20.0) is placed in a 2.00 L flask at 50°C, what is/are the physical state(s) of the compound in the flask. Show by calculation.
6. Using the boiling point data and the vapor pressure of the compound at 50°C, what is ΔH_{vap} of the compound?

C. Calcium has an atomic radius of 0.197 nm and its density is 1.54 g/cm^3. Is the calcium unit cell simple cubic, body-centered cubic or face-centered cubic?
(*Hint:* In a simple cubic cell, there is 1 atom/cell. There are 2 atoms/cell in a body-centered cubic cell and 4 atoms/cell in a face-centered cell.)
You must support your answer with calculations.

Liquids and Solids

Worksheet B

A. Answer the following questions:
1. Consider calcium carbonate, $CaCO_3$.
   a. What kind of compound is it? (network covalent, molecular, ionic, metal)
   b. What kinds of bonds are present in the solid?
   c. When calcium carbonate is converted by heat to CO_2 and CaO (s), what bonds are broken?
   d. Compare its melting point to $C_{diamond}$ and CO_2.
   e. What types of bonds are broken when calcium carbonate is dissolved in water?

2. Consider the following data given for carbon tetrafluoride, CF_4, and phosphorus trifluoride, PF_3.

Compound	$\mathcal{M}$ (g/mol)	No. of e^-	Melting Pt (°C)	Boiling Pt (°C)
CF_4	88	42	− 184	− 128
PF_3	88	42	− 152	− 102

Explain why CF_4 has a lower melting point and boiling point than PF_3.

3. Consider methyl iodide, CH_3I. It is a liquid at room temperature ($\approx 20°C$), and 1 atm pressure. Radioactive methyl iodide can be prepared by using radioactive I–131 as the iodine source. In an experiment using the apparatus illustrated below, nonradioactive methyl iodide is put in the left side of the glass partition, while radioactive CH_3I is put on the right side. The apparatus is tightly covered. After a day, the liquid on the left side is tested for radioactivity.

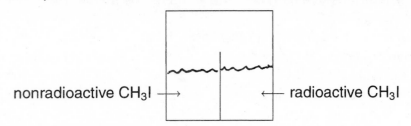

nonradioactive CH_3I $\longrightarrow$ $\longleftarrow$ radioactive CH_3I

   a. What result would you expect?
   b. Explain your reasoning for the answer to (a).
   c. How would you modify the apparatus so that the nonradioactive CH_3I does not get radioactive?

B.   Consider the phase diagram for compound Z given below.
1.   What phase(s) is/are present at point X?
2.   What is the triple point for Z?
3.   Can the compound be liquified at 150°C by increasing the pressure to 830 mm Hg?
4.   Can the compound be liquified at 100°C by increasing the pressure to 800 mm Hg?
5.   What phase(s) is/are present at 35°C and 730 mm Hg?
6.   What is the normal boiling point of the compound?
7.   What pressure is required to make the compound boil at 75°C?
8.   What process is indicated by the arrow ⟶   ?
9.   What process is indicated by the arrow ⟿⟶   ?
10. Which phase is the denser phase?

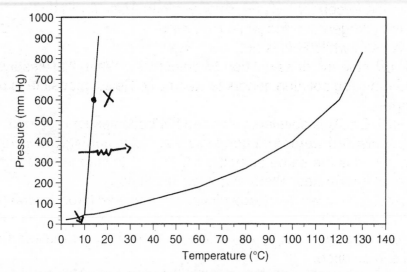

C.   A compound Y, has a normal boiling point of 78°C and a vapor pressure of 122 mm Hg at 54°C. Answer (by calculation) the following questions about the compound.
1.   If 0.200 moles of compound Y are put into a 30.0 L flask at 54°C, what phase(s) of Y are in the flask?
2.   If your answer in (1) was all gas, what is the maximum volume of flask that can be used to have both gas and liquid (of the same amount of Y) present in the flask at 54°C?
     If your answer in (1) was gas and liquid, what volume of flask would be needed to have only gas (of the same amount of Y) present in the flask at 54°C?
3.   What is ΔH_{vap} for Y?
4.   At what temperature will Y have a vapor pressure of 368 mm Hg?

Liquids and Solids

Worksheet C

A. Circle the correct answer(s).

1. The boiling point of ammonia (NH_3) is –33°C, while the boiling point of phosphine (PH_3) is –87.7°C. Ammonia has a higher boiling point because
   a. the molar mass of NH_3 is less than that of PH_3.
   b. dispersion forces for NH_3 are stronger than those for PH_3.
   c. NH_3 is hydrogen bonded while PH_3 is not.
   d. NH_3 is polar while PH_3 is not.

2. Solid calcium chloride does not conduct electricity. When it is dissolved in distilled water, the resulting solution conducts electricity. Select plausible explanations for this change.
   a. Like HCl, $CaCl_2$ is covalently bonded but becomes ionic when dissolved.
   b. It is the distilled water that conducts electricity, adding $CaCl_2$ has nothing to do with it. It is the same as being careful not to step in water when handling electrical appliances. Water conducts electricity.
   c. Since calcium loses $2e^-$ becoming Ca^{2+}, these electrons moving in the water conduct electricity.
   d. The ions in solid $CaCl_2$ are not free to move but are released and become mobile in solution.
   e. All compounds with chlorine atoms have that property, non-conductor when solid, conductor when dissolved.

3. The normal boiling point of a liquid
   a. is the temperature at which liquid and vapor are in equilibrium.
   b. is the temperature at which the vapor pressure is 1.00 atm.
   c. varies with atmospheric pressure.
   d. is the temperature at which the vapor pressure equals external pressure.
   e. can be influenced by H–bonding.

4. In general, which of the following properties are similar for network covalent and nonpolar molecular solids?
   a. high melting points
   b. nonelectrical conductors
   c. insoluble in water
   d. good thermal conductor
   e. presence of covalent bonds

5. Consider the following phase diagram. Circle the correct observations based on it.

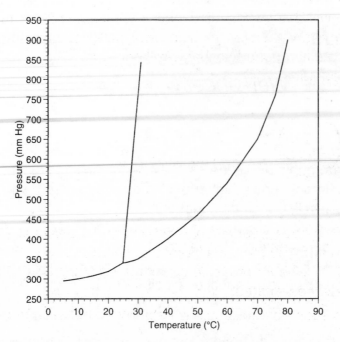

a. At 25°C and 340 mm Hg, solid, liquid, and vapor are in equilibrium.
b. The normal boiling point is close to 70°C.
c. As pressure is applied to the solid, the melting point decreases.
d. The temperature at which sublimation occurs decreases as the external pressure on the solid is decreased.
e. The density of the liquid phase is larger than the density of the solid phase.
f. It is impossible to obtain this compound in the liquid state at 100°C and 1000 mm Hg.

B. Consider the vapor pressure curves of molecules and answer the following questions.

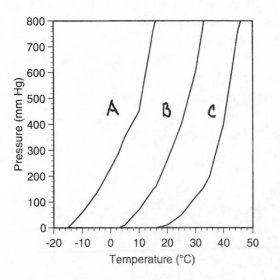

1. Which compound has the strongest forces between molecules?
2. Which has a normal boiling point of ≈ 15°C?
3. At what temperature will B boil if the atmospheric pressure is 500 mm Hg?
4. At 25°C and 400 mm Hg, what is(are) the physical states of each compound?
5. What external pressure is required so that C will boil at 40°C?

C. Consider the following apparatus:

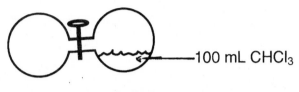

Bulb A   Bulb B

— Both bulbs A and B each have a volume of 500.0 mL and are kept at 20°C.
— Bulb B was evacuated and 100.0 mL of chloroform, $CHCl_3$ ($\mathcal{M}$ = 119.4 g/mol, d = 1.478 g/mL) is added. The vapor pressure of chloroform at 20°C is 159.6 mm Hg.
— Bulb A is also completely evacuated.

The valve is opened. As soon as equilibrium is established between the two valves, the valve is closed.

Answer the following questions (show calculations). Neglect the volume decrease of the liquid in calculating the volume of the gas phase after the valve is opened and the volume of the tube connecting both bulbs.

1. What is the volume of liquid chloroform in bulb B?
2. What is the pressure in both bulbs after the valve is closed when equilibrium is established?
3. What is the mass of chloroform gas in bulb A?
4. What is the mass of chloroform gas in bulb B?

Liquids and Solids

Answers

Worksheet A

A. 1. d, f 　　　　　　2. d 　　　　　　3. a, c, d

B. 1.

A = triple point
AB = melting point curve
AC = vapor pressure curve
AD = sublimation curve

　　2. solid phase is denser 　　　3. vaporization 　　　4. freezing
　　5. liquid and gas 　　　　　　6. 26.5 kJ/mol

C. face-centered cubic cell

Worksheet B

A. 1. a. ionic
　　　b. ionic and covalent
　　　c. ionic and covalent
　　　d. melting point $C_{diamond}$ > melting point $CaCO_3$
　　　　 melting point CO_2 < melting point $CaCO_3$
　　　e. ionic bonds
　2. PF_3 is polar, CF_4 is not.
　3. a. nonradioactive CH_3I has become radioactive.
　　　b. Dynamic equilibrium – vapor phase contains both types of molecules.
　　　　 Radioactive molecules can condense into nonradioactive liquid.
　　　c. Make a partition that goes all the way to the top of the container.

B. 1. solid and liquid 　　　　2. 45 mm, 10°C 　　　　3. no
　4. yes 　　　　　　　　　　5. liquid 　　　　　　　6. ≈ 125°C
　7. ≈ 220 mm Hg 　　　　　 8. sublimation 　　　　9. melting
　10. solid

C. 1. liquid and gas
   2. If answer is all gas – V must be 33 L but the flask is only 30 L, so the answer is wrong.
      If answer is liquid and gas – V must be less than 33 L, answer is correct.
   3. 72.7 kJ/mol
   4. 341 K

Worksheet C

A. 1. c   2. d   3. b, e   4. b, c, e   5. a, b, d, f

B. 1. C
   2. A
   3. ≈ 30°C
   4. A – all gas;   B – vapor/liquid at equilibrium;   C – mostly liquid
   5. ≈ 400 mm Hg

C. 1. 99.37 mL
   2. 159.6 mm Hg
   3. 0.521 g
   4. 0.417 g

Solutions

Worksheet A

A. Circle the correct answer(s).

1. The solubility of an ionic compound in water is 11 g/100 g H_2O at 25°C. A solution which contains 15 g of the compound and 250 g of water would
    a. be saturated.
    b. be unsaturated.
    c. be an electrolyte.
    d. lower the freezing point of the solvent.
    e. not dissolve if the pressure over the solution is decreased.

2. Which of the following solutions CANNOT be prepared by making only mass measurements?
    a. X_{NaCl} = 0.050
    b. 1 M NaCl
    c. 12.1% (by volume) NaCl
    d. 1 m NaCl
    e. a saturated solution of NaCl

3. Consider the solution of a solid in water.
    $$X(s) \rightarrow X(aq) \qquad \Delta H = -75 \, kJ$$
    a. Heating the solution increases solubility.
    b. Increasing the pressure over the solution increases the solubility.
    c. The solid is not ionic.
    d. The solution feels warmer after the solid is dissolved.
    e. The solid is an electrolyte.

4. If the pressure of a gas over a liquid is doubled, the concentration of gas dissolved in the liquid will
    a. stay the same.   b. be doubled.
    c. be halved.   d. increase by 50%.
    e. increase by 100%.

5. Consider 3 tubes. Tube A has pure water. Tube B has an aqueous 1.0 m solution of ethanol, C_2H_5OH. Tube C has an aqueous 1.0 m solution of NaCl. Assume that both NaCl and C_2H_5OH have a concentration of 1.0 M.
    a. The vapor pressure of solvent over tube A is greater than the vapor pressure of solvent over tube B.
    b. The freezing point of tube B is higher than the freezing point of tube A.
    c. The freezing point of tube C is higher than the freezing point of tube A.
    d. The boiling point of tube B is higher than the boiling point of tube C.
    e. The osmotic pressure of tube B is greater than the osmotic pressure of tube C.

B. Consider an aqueous solution of sugar, $C_{12}H_{22}O_{11}$ (M = 342 g/mol), in water. Its density is 1.05 g/mL. The mole fraction of sugar in solution is 0.0286. Calculate
    1. the vapor pressure of the solvent over the solution at 25°C. The vapor pressure of pure water at 25°C is 23.8 mm Hg.

    2. the boiling point of the solution.

    3. the freezing point of the solution.

    4. the osmotic pressure of the solution at 25°C.

C. A solution is made up of 5.45 g of an unknown solid in 100.0 g of bromoform (k_f = 14.4 °C/m). The solution has a freezing point 4.32°C lower than the freezing point of bromoform. What is the molar mass of the unknown solid?

Solutions

Worksheet B

A. Answer the following questions.

1. Consider aqueous solutions prepared by adding 0.1 mol of the following solutes to 1.00 kg of water.

   $C_6H_{12}O_5$, C_2H_5OH , NaI , $Mg(NO_3)_2$, $Al(NO_3)_3$

   a. Which solution(s) has(have) the highest boiling point?
   b. Which solution(s) has(have) the lowest boiling point?
   c. Which solution(s) has(have) the highest freezing point?
   d. Which solution(s) has(have) the lowest freezing point?
   e. Assuming that the molarity of the solutions are the same as their molality, which solution has the highest osmotic pressure?

2. Beaker A has 1.00 kg of water and 0.100 mol of glucose. Beaker B has 1.00 kg of water and 0.100 mol of $CaCl_2$. Answer the questions below, using **LT** *(for less than)*, **GT** *(for greater than)*, **EQ** *(for equal to)*, or **MI** *(for more information required)* in the blanks provided.

   a. The vapor pressure of pure water _____ the vapor pressure of the solution in Beaker A.

   b. The boiling point of the solution in Beaker A _____ boiling point of the solution in Beaker B.

   c. The molarity of the solution in Beaker A _____ molarity of the solution in Beaker B.

   d. The freezing point of the solution in Beaker A _____ freezing point of the solution in Beaker B.

   e. The solubility of the solute in Beaker A _____ solubility of the solute in in Beaker B.

3. Which of the following phenomena is best described by Henry's Law?
    a. Increasing temperature favors an endothermic process.
    b. The solubility of $NaCl(s)$ in water increases with increasing temperature.
    c. The solubility of $MgCl_2(s)$ in water is unaffected by pressure.
    d. The vapor pressure of an aqueous solution of glucose, $C_6H_{12}O_6$, is lower than the vapor pressure of pure water.
    e. The solubility of $O_2(g)$ in water increases with increasing pressure.

B. What is the boiling point of an aqueous solution of a nonelectrolyte that has a freezing point of $-2.86°C$?

C. An aqueous solution is both 0.200 m and 0.227 M. The solute has a molar mass of 115 g/mol.
    1. What is the density of the solution?

    2. What is the mass percent of solute in solution?

Solutions

Worksheet C

A. State whether the statement is true or false. If it is false, rewrite it to make it true.
   1. HF is more soluble than $CHCl_3$ in water.
   2. The solubility of CO_2 (g) in water increases with increasing temperature.
   3. The solubility of CaO (s) in water increases with increasing pressure.
   4. Without more information, one cannot know the effect of increasing temperature on the solubility of $Zn(NO_3)_2$ (s) in water.
   5. A solution of urea, $(NH_2)_2CO$, is nonconducting.
   6. The vapor pressure of the solvent above a solution is always lower than the vapor pressure of the pure solvent.
   7. The osmotic pressure of a 0.10 M aqueous solution of NaCl is almost half as much as that of a 0.10 M solution of sucrose, $C_{12}H_{22}O_{11}$.
   8. To convert molarity to molality, the density of the solution must be given.
   9. A solution is prepared by dissolving 15.0 g of solute in 275 g of water. The solubility of the solute in water is 5.00 g/100 g H_2O. The resulting solution is saturated.
   10. A given volume of a 1.0 M solution is mixed with an equal volume of water. In the resulting diluted solution, the number of moles of solute remains the same.
   11. If the solute–solvent interactions are strong, the solute will not dissolve in the solvent.
   12. The most important factor in determining the solubility of a solute in a solvent is the similarity in intramolecular forces (forces within the units).
   13. According to Raoult's law, the vapor pressure of a solvent over a solution increases as the mole fraction of solute in the solution increases.
   14. Dissolving CO_2 (g) in water is an exothermic process. The solubility of CO_2 (g) in a carbonated beverage pressurized to 3 atm at 25°C is 9.4×10^{-2} M. Its solubility at 12.5°C and 3 atm would be 0.19 M.
   15. A 0.10 m aqueous solution of an electrolyte will have a greater effect on the colligative properties of the solution than a 0.10 m aqueous solution of a nonelectrolyte.

B. Rank the following solutes: $H_3C - O - CH_3$, NaCl, C_4H_9OH
   1. with respect to increasing solubility in water.
   2. as aqueous solutions (0.10 m) with respect to increasing freezing point.
   3. as aqueous solutions (0.10 m) with respect to increasing boiling point.
   4. as aqueous solutions (0.10 m) with respect to increasing vapor pressure of the solvent over the solution.
   5. as aqueous solutions (0.10 M) with respect to increasing osmotic pressure.

C. A solution is prepared by mixing 345 g of a nonvolatile solute in 250.0 mL of methanol, CH_3OH, at 50°C. The resulting solution has a vapor pressure of 0.376 atm at 50°C. Methanol has a density of 0.796 g/mL, a molar mass of 32.04 g/mol, and a vapor pressure of 0.526 atm at 50°C. What is the molar mass of the solute?

D. A 10.0 m aqueous solution of NaOH has a density of 1.30 g/mL at room temperature. For the solution, calculate
    1. the mole fraction of NaOH.

    2. its molarity.

    3. the mass percent of solute.

Solutions

Answers

Worksheet A

A. 1. b, c, d   2. b, c   3. c, d   4. b, e   5. a

B. 1. 23.1 mm Hg   2. 100.85°C   3. −3.03°C   4. 26.9 atm

C. 182 g/mol

Worksheet B

A. 1. a. $Al(NO_3)_3$ b. $C_6H_{12}O_6$, C_2H_5OH c. $C_6H_{12}O_6$, C_2H_5OH
     d. $Al(NO_3)_3$ e. $Al(NO_3)_3$

  2. a. GT   b. LT   c. MI   d. GT   e. MI

  3. e

B. 100.80°C

C. 1. 1.16 g/mL   2. 2.25%

Worksheet C

A. 1. True
   2. False — decreases with increasing T
   3. False — Pressure has no effect on the solubility of CaO in water.
   4. True
   5. True
   6. True
   7. False — almost twice as much
   8. True
   9. False — supersaturated
   10. True
   11. False — will dissolve
   12. False — intermolecular (forces between the units)
   13. False — decreases
   14. False — The solubility would be greater than 9.4×10^{-2} M, but there is not enough information to know the exact solubility at that temperature.
   15. True

B. 1. $H_3C - O - CH_3 < C_4H_9OH < NaCl$ 2. $NaCl < (H_3C - O - CH_3 = C_4H_9OH)$
   3. $(H_3C - O - CH_3 = C_4H_9OH) < NaCl$ 4. $(H_3C - O - CH_3 = C_4H_9OH) < NaCl$
   5. $(H_3C - O - CH_3 = C_4H_9OH) < NaCl$

C. 139 g/mol

D. 1. $X_{NaOH} = 0.153$ 2. M = 9.26   3. 28.6% NaOH

Rate of Reaction

Worksheet A

A. Circle the true statement(s).
1. The presence of a catalyst
   a. increases the rate of a reaction.
   b. should decrease the rate of a reaction.
   c. does not change E_a (i.e., $E_{a_{uncat}} = E_{a_{cat}}$).
   d. changes k (i.e., $k_{uncat} \neq k_{cat}$).

2. For a zero-order reaction
   a. the reaction concentration does not change with time.
   b. the reaction rate decreases linearly with time.
   c. the reaction rate is constant.
   d. units for k are mole/L-time.

3. For a second order reaction
   a. k has units of mol^2/L^2-time.
   b. $1/[X]$ vs time is a linear plot.
   c. the half-life is dependent on initial concentration.
   d. the rate quadruples when the concentration is doubled.

4. For a first order reaction
   a. the rate is directly proportional to concentration so that when concentration is doubled, the rate is also doubled.
   b. the function relating $\ln[X]$ to time is decreasing.
   c. a linear plot is obtained when plotting $\ln[X]$ vs rate.
   d. the unit for k is reciprocal time ($time^{-1}$).

5. The dimerization of NO_2 is a second order reaction.

   $$2\,NO_2\,(g) \;\rightarrow\; N_2O_4\,(g)$$

   The rate expression for the dimerization is
   $$rate = k[NO_2]^2$$
   For this reaction
   a. k changes when pressure is doubled.
   b. rate doubles when the volume of the container is doubled but k remains the same.
   c. adding more NO_2 increases rate and k.
   d. increasing the temperature increases the rate and k.
   e. decreasing the concentration of NO_2 decreases the rate but k remains constant.

B. Consider the following energy diagram (not to scale) for the reaction

$$2\,CH_3\,(g) \;\rightarrow\; C_2H_6\,(g)$$

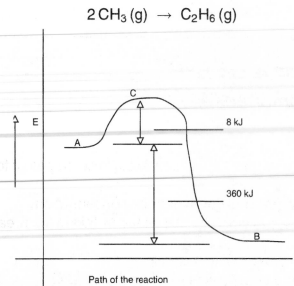

Path of the reaction

1. What is the activation energy of the forward reaction?
2. At what point in the diagram would $2\,CH_3$ (g) appear?
3. What is ΔH for the reaction?
4. What is the activation energy for the reverse reaction?
5. At what point in the diagram would C_2H_6 be certainly found?
6. What is any species at point C called?

C. For the reaction

$$2\,A\,(g) \;\rightarrow\; products \qquad\qquad \Delta H = 22\;kJ$$

The following experimental data is obtained at 25°C.

[A]	rate (mol/L-min)
0.200	0.0123
0.300	0.0185
0.500	0.0308

For this reaction:

1. Calculate the half-life when [A] = 0.400 M.

2. How long will it take to decompose 38% of A at 25°C?

3. How fast is the decomposition at 25°C when [A] = 0.750 M?

4. What is E_a if the rate constant doubles when the temperature at which the experiment is performed is increased by 10°C?

Rate of Reaction

Worksheet B

A. Mark the statements below as true or false. If the statement is false, restate it to make it true.

1. In chemical kinetics, the rate constant is dependent on concentration.
2. Graphically, the instantaneous rate at time t is determined by taking the area under the curve.
3. A large rate constant means that the reaction is fast.
4. The half-life of any reaction is the time it takes to make $[X_o] = \frac{1}{2}[X]$.
5. The rate of a catalyzed reaction is independent of the concentration of the catalyst.
6. All collisions between atoms, molecules or ions will result in a reaction if the reactant particles have energies equal to or greater than the activation energy.
7. For a reaction that occurs in a series of successive steps, the rate of the reaction depends on the slowest step.
8. For the reaction

$$A + B \rightarrow C + D \qquad \Delta H = 25 \text{ kJ}$$

The activation energy, E_a, is at least 25 kJ, but the precise value cannot be determined from the information given with the reaction.
9. In a zero-order reaction, the reactant's concentration remains the same throughout the reaction.
10. The units for rate, regardless of order, are always mole/L-time.
11. The order of a reaction can only be determined experimentally.
12. The rate constant for any order reaction increases when the initial concentration of the reactants are increased.

The next three statements are about the reaction

$$2A + B \rightarrow C + 2D$$

The reaction is first order in A and zero order in B.

13. The rate expression for the reaction is rate = $k[A]^2[B]$
14. A plot of ln [A] vs time gives a straight line.
15. The half-life of B is independent of its initial concentration.

B. Consider the graph about the following reaction at 25°C (line A), and answer the questions below.

X → products

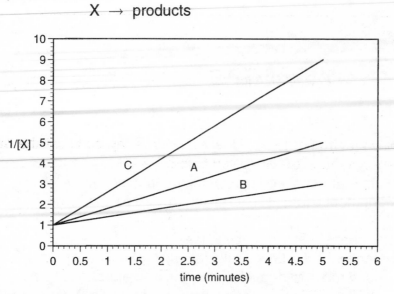

1. What is the rate constant for the reaction at 25°C?
2. What is the order of the reaction?
3. Write the rate expression for the reaction.
4. What is the half-life of the reaction at 25°C, when [X] = 0.500 M?
5. How fast is the reaction at 25°C going, when [X] = 0.200 M?
6. Which line could represent a plot of the same reaction at 15°C? at 45°C?
7. If B represents the reaction at 25°C but with a catalyst, what effect does the catalyst have on the reaction?
8. If C represents the reaction at 25°C but with a catalyst, what effect does the catalyst have on the reaction?

C. Given the following data for the reaction

A + B → products

and assuming that the order of each reactant does not change with an increase in temperature, answer the following questions.

Expt.	Temp	[A]	[B]	rate (mol/L-hr)
1	25°C	0.100 M	0.100 M	0.0375
2	25°C	0.100 M	0.500 M	0.0375
3	25°C	0.200 M	0.100 M	0.0750
4	45°C	0.100 M	0.200 M	0.1120

1. How long will it take to use up half of A?
2. At what concentrations of A and B will the reaction rate be 0.186 mol/L-hr?
3. What percent of A will be used up after 25 minutes?
4. What is the activation energy of the reaction?

Rate of Reaction

Worksheet C

A. Answer the following questions about reactions X and Y. They occur at the same temperature and their energy diagrams are given below.

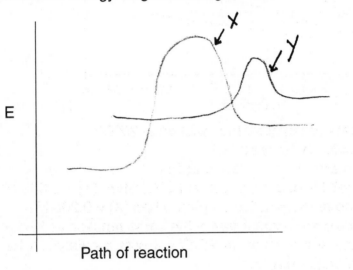

1. Which reaction is exothermic?
2. Which reaction has a lower activation energy?
3. Which reaction has a smaller k?
4. Which reaction is faster (assuming identical initial concentrations)?
5. Which reaction will increase the value of k when the temperature is increased?

B. For the reaction

$$N_2O_5 \text{ (g)} \rightarrow 2\,NO_2 \text{ (g)} + \tfrac{1}{2}\,O_2 \text{ (g)} \qquad \Delta H° = 55.2 \text{ kJ}$$

k is $8.0 \times 10^{-7}\,s^{-1}$ at 0.0°C and $8.9 \times 10^{-4}\,s^{-1}$ at 50.0°C.

1. Calculate the activation energy of the reaction.
2. Draw an energy diagram for the reaction.
3. What is the order of the reaction for the decomposition of N_2O_5?
4. How long will it take to decompose 0.100 M N_2O_5 to 0.0500 M at 50°C?
5. What is the rate of the reaction when the initial concentration of N_2O_5 is 2.65 M at 50°C?

C. For the reaction

   $$A + 2B \rightarrow products$$

the following data is obtained.

Expt.	[A]	[B]	rate (mol/L-s)
1	0.0800 M	0.0400 M	0.440
2	0.0800 M	0.0800 M	0.621
3	0.1600 M	0.0400 M	1.760

1. What is the rate expression for the reaction?
2. What is the rate constant for the reaction?
3. What is the rate of the reaction when [A] = 0.300 M and [B] = 2[A]?

Rate of Reaction

Answers

Worksheet A

A. 1. a, d   2. c, d   3. b, c, d   4. a, b, d   5. d, e

B. 1. 8 kJ   2. A   3. −360 kJ   4. 368 kJ   5. B
   6. activated complex

C. 1. 11.3 min   2. 7.8 min   3. 0.0461 mol/L-min   4. 52.9 kJ

Worksheet B

A. 1. False — rate is dependent on concentration
   2. False — by taking the slope of the tangent to the curve at point t.
   3. True
   4. False — $\frac{1}{2}[X_o] = [X]$
   5. False — dependent on [catalyst]
   6. False — add: and are oriented to make an effective collision.
   7. True
   8. True
   9. False — the rate
   10. True
   11. True
   12. False — temperature
   13. False — rate = k[A]
   14. True
   15. False — half life of A   *or*   is $[A_o]/2k$

B. 1. 0.77 L/mol-min   2. 2nd
   3. rate = k $[X]^2$ 4. 2.6 min
   5. 0.031 mol/L-min   6. Line B at 15°C; line C at 45°C
   7. Catalyst slows down the reaction.   8. Catalyst speeds up the reaction.

C. 1. 1.85 hr   2. [A] = 0.496 M ; [B] = any concentration
   3. 14%   4. 43.1 kJ

Worksheet C

A. 1. reaction X   2. reaction Y   3. reaction X
   4. reaction Y   5. both reactions

B. 1. 103 kJ
   2.

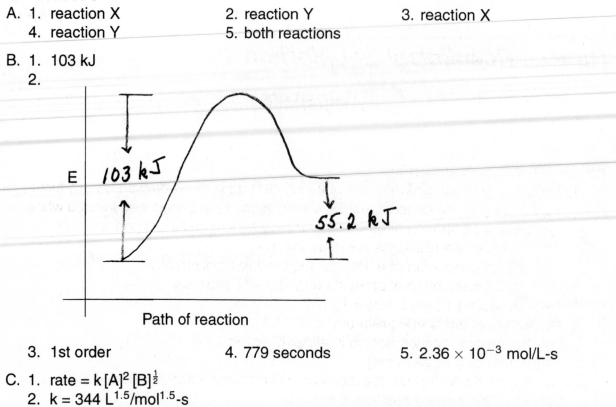

Path of reaction

   3. 1st order   4. 779 seconds   5. 2.36×10^{-3} mol/L-s

C. 1. rate = $k\,[A]^2\,[B]^{\frac{1}{2}}$
   2. $k = 344$ $L^{1.5}/mol^{1.5}$-s
   3. rate = 24.0 mol/L-s

Gaseous Chemical Equilibrium

Worksheet A

A. Circle the correct answer(s).

1. When a graph is drawn which plots concentration of reactants and products (y–axis) versus time (x–axis), equilibrium is seen to have been established when
   a. the concentration of products and reactants remains unchanged.
   b. 100% of the reactants are consumed.
   c. the concentration of reactants and products are equal.
   d. the concentration of products only remains constant.
   e. [products] + [reactants] = 1

2. When a system is at equilibrium
   a. no further reaction occurs in either direction.
   b. [products] = [reactants]
   c. forward and reverse reactions are taking place simultaneously.
   d. k for the forward reaction is equal to k for the reverse reaction.
   e. the rate of both forward and reverse reactions is the same.

3. A chemical reaction has reached equilibrium when
   a. all reactions stop.
   b. the equilibrium constant is 1.
   c. the partial pressures meet the requirements of the equilibrium constant expression.
   d. Q = K
   e. all the reactants are converted to products.

4. Consider the system

$$CaCO_3\,(s) \;\rightleftharpoons\; CaO\,(s) \;+\; CO_2\,(g)$$

   After the system reaches equilibrium, [CaO] is doubled. What is the result of this change?
   a. K changes.
   b. $(P_{CO_2})_{eq}$ increases.
   c. $(P_{CO_2})_{eq}$ decreases.
   d. $(P_{CO_2})_{eq}$ remains the same.

5. Which of the following statements are true?
   a. Q and K can sometimes have the same numerical value.
   b. Q can sometimes be 0.
   c. Q can sometimes be less than K.
   d. The numerical value for Q changes with time as the reaction proceeds.

6. For the system

$$NH_4Cl\,(g) \;\rightleftharpoons\; NH_3\,(g) \;+\; HCl\,(g)$$

If the pressure of NH_3 is doubled (by compression) <u>after</u> the system reaches equilibrium, K will

a. double.

b. increase but by less than a factor of 2.

c. be halved.

d. decrease but by less than a factor of 2.

e. do none of the above.

B. Consider the system

$$N_2\,(g) \;+\; 3\,H_2\,(g) \;\rightleftharpoons\; 2\,NH_3\,(g)$$

A 2.568–g sample of NH_3 was placed in a 2.000 liter flask at 25°C. When equilibrium was reached at that temperature, it was determined that the ammonia was reduced to 75.0% of its original value.

1. Calculate K for the decomposition of 2.00 moles of ammonia at 25.0°C.

2. Calculate K for the decomposition of 2.00 moles of ammonia at 50°C. (ΔH_f° for NH_3 is –46.1 kJ/mol)

C. For the system

$$PCl_5\,(g) \;\rightleftharpoons\; PCl_3\,(g) \;+\; Cl_2\,(g)$$

K is 1.0 at a certain temperature. If one starts with 0.50 atm of both PCl_3 and Cl_2 (g)

1. In what direction will the reaction proceed?

2. What are the partial pressures of all species when equilibrium is reached?

3. If after equilibrium is reached, the partial pressure of chlorine gas is reduced to 0.2 atm, what are the partial pressures of all species when equilibrium is re-established?

4. When stress is imposed on the system, does the system fully recover from that stress (i.e., Are the equilibrium pressures the same before and after the stress?)?

Gaseous Chemical Equilibrium

Worksheet B

A. Circle the correct answer(s).
1. Consider the system

$$CaCO_3\,(s) \rightleftharpoons CaO\,(s) + CO_2\,(g)$$

After the system reaches equilibrium, [CaO] is doubled. What is the result of this change?
a. K changes.
b. $(P_{CO_2})_{eq}$ increases.
c. $(P_{CO_2})_{eq}$ decreases.
d. $(P_{CO_2})_{eq}$ remains the same.
e. Nothing, because no reaction in either reaction is taking place.

2. For the system

$$NH_4Cl\,(g) \rightleftharpoons NH_3\,(g) + HCl\,(g)$$

The pressure is doubled by compression after the system reaches equilibrium. When equilibrium is reestablished,
a. $(P_{NH_4Cl})_{eq}$ increases.
b. $(P_{NH_3})_{eq}$ and $(P_{HCl})_{eq}$ increases.
c. the value of K doubles.
d. the value of K increases, but by less than a factor of 2.
e. the pressure of NH_4Cl is halved.

3. For the reaction

$$N_2\,(g) + O_2\,(g) \rightleftharpoons 2\,NO\,(g)$$

K is 1.1×10^{-3} at 1900°C and 3.6×10^{-3} at 2200°C.
a. The reaction is exothermic.
b. An increase in temperature will shift the direction of the reaction so that the partial pressures of N_2 and O_2 will increase.
c. At either temperature, compression will not change K, but will change the equilibrium pressures of N_2, O_2, and NO.
d. At 1000°C, one can expect K to be less than 1.1×10^{-3}.
e. Adding argon gas will shift equilibrium to the right.

4. At the point of chemical equilibrium,
   a. the rates of the forward and reverse reactions are zero.
   b. the rate of the forward reaction is equal to the rate of the reverse reaction.
   c. $K = 1$.
   d. $Q = K$.
   e. the rate constant for the forward reaction (k_1) is equal to the rate constant for the reverse reaction (k_{-1}).

5. When the value of K for a reaction is small (< 0.10), it means that
   a. the partial pressures of the reactants are small.
   b. the partial pressures of the products are small with respect to the reactants.
   c. the reaction in the forward direction ($\longrightarrow$) is slow.
   d. the temperature needs to be increased so the value of K is increased.
   e. the volume of the container needs to be increased to increase the partial pressures of the products.

B. Consider the system

$$N_2\,(g) + 3\,H_2\,(g) \;\rightleftharpoons\; 2\,NH_3\,(g) \qquad\qquad \Delta H = -92.2\ kJ$$

Show the effect on the partial pressure of H_2 and on the numerical value of K by using $\uparrow$ for an increase, $\downarrow$ for a decrease, and NC for no change.

	P_{H_2}	K
1. adding NH_3	_____	_____
2. doubling the volume	_____	_____
3. increasing T	_____	_____
4. adding Ar (g)	_____	_____
5. adding N_2 (g)	_____	_____

C. For the decomposition of $CaCO_3$ at 850°C:

$$CaCO_3\,(s) \rightleftharpoons CaO\,(s) + O_2\,(g) \qquad\qquad K = 0.49$$

What percent by mass of $CaCO_3$ is converted to CaO if one mole of $CaCO_3$ is heated to 8.50×10^{2}°C in a 10.0 L evacuated flask and the reaction is allowed to reach equilibrium?

D. For the system

$$A\,(g) \rightleftharpoons B\,(g) + C\,(g) \qquad\qquad \Delta H = -12.0\ kJ$$

Equilibrium partial pressures are 1.2 atm, 2.0 atm, and 2.2 atm, for A, B, and C, respectively, at 37°C in a 10.0 L flask.
1. Calculate K for the system at 37°C.
2. If the flask is expanded to 20.0 L, what are the partial pressures when equilibrium is re-established?
3. What is K at 89°C?

Gaseous Chemical Equilibrium

Worksheet C

A. Write T if the statement is true, and F if the statement is false.

_____ 1. In all chemical reactions that go to completion, $K = 1$.

_____ 2. $K = \dfrac{\text{rate forward reaction}}{\text{rate reverse reaction}}$

_____ 3. At equilibrium, $\dfrac{\text{rate forward reaction}}{\text{rate reverse reaction}} = 1$

_____ 4. At equilibrium, the partial pressures of both products and reactants can be, but need not be, equal.

_____ 5. In order to achieve equilibrium, all substances involved must be initially present.

_____ 6. The equilibrium constant K, always increases with an increase in temperature.

_____ 7. The reaction quotient Q and the equilibrium constant, K, always have the same numerical value.

_____ 8. The numerical value of Q, when compared to K, indicates how fast the forward reaction is.

_____ 9. A chemical reaction with a large equilibrium constant $(K > 10)$ always reaches equilibrium rapidly.

_____ 10. A small equilibrium constant $(K < 0.10)$ means that at equilibrium, one can expect to find more of the species on the left side of the equation than those on the right side of the equation.

_____ 11. If at equilibrium $P_A > P_B$, then at equilibrium $n_A > n_B$.

_____ 12. For a system at equilibrium, where k_1 is the rate constant for the forward reaction and k_{-1} is the rate constant for the reverse reaction, $k_1 \neq k_{-1}$.

_____ 13. K is always positive.

_____ 14. When $Q < K$, the reaction shifts to the right and all the reactants are converted to products.

_____ 15. When equilibrium is reached, both reactions (forward and reverse) stop. The reactions resume when a stress is imposed on the system.

B. Fill in the blank with the appropriate symbol. (↑ for increase, ↓ for decrease, and NC for stays the same.)
    1. An evacuated flask with a piston contains $CaCO_3$. The flask is heated and $CaCO_3$ decomposes according to the equation

    $$CaCO_3 (s) \rightarrow CaO (s) + CO_2 (g)$$

    At equilibrium, the flask contains $CaCO_3$, CaO, and CO_2. The volume of the flask is doubled by raising the piston. When equilibrium has been re-established, the flask still has $CaCO_3$, CaO, and CO_2. After equilibrium is re-established:

    a. The mass of CO_2 _____ .

    b. The partial pressure of CO_2 _____ .

    c. The mass of $CaCO_3$ _____ .

    2. The decomposition of NH_4Br at 400°C is endothermic.

    $$NH_4Br (s) \rightleftharpoons NH_3 (g) + HBr (g)$$

    After equilibrium is established, the reaction vessel contains NH_4Br, NH_3, and HBr.

    a. P_{HBr} _____ when extra NH_3 is added.

    b. n_{NH_4Br} _____ when the volume is doubled.

    c. P_{NH_3} _____ when the mass of NH_4Br is increased.

    d. When T is increased, P_{NH_3} _____ , n_{NH_4Br} _____ , and K _____ .

C. For the reaction

    $$2 NaHCO_3 (s) \rightleftharpoons Na_2CO_3 (s) + CO_2 (g) + H_2O (g)$$

    K is 0.23 at 100°C.
    1. What is the total pressure at equilibrium?
    2. If one starts with 0.500 mol of $NaHCO_3$ at 100°C in a 10.0 L evacuated flask, can equilibrium be established?
    3. If your answer to (2) is yes, what are the partial pressures of CO_2 and H_2O at equilibrium? What are the masses of $NaHCO_3$ and Na_2CO_3?

D. For the reaction

    $$2 A (g) + B (g) \rightleftharpoons C (g) + D (s) \qquad \Delta H = -33.0 \text{ kJ}$$

    1. Initially in a 5.0 L flask at 75°C, there are 0.20 mol A, 0.30 mol C, and 1.00 mol D. When equilibrium is established, $P_C = 1.0$ atm. What are the partial pressures of all the gases at equilibrium?
    2. How many moles of D are present at equilibrium?
    3. At what temperature will the equilibrium constant, K, be 1?

Gaseous Chemical Equilibrium

Answers

Worksheet A
A. 1. a   2. c, e   3. c, d   4. d   5. a, b, c, d   6. e

B. 1. 0.040   2. 0.71

C. 1. $\longleftarrow$
   2. $P_{PCl_5} = 0.13$ atm; $P_{PCl_3} = P_{Cl_2} = 0.37$ atm
   3. $P_{PCl_5} = 0.10$ atm; $P_{PCl_3} = 0.40$ atm; $P_{Cl_2} = 0.23$ atm
   4. No

Worksheet B
A. 1. d   2. a   3. c, d   4. b, d   5. b

B. 1. $\uparrow$; NC   2. $\uparrow$; NC   3. $\uparrow$; $\downarrow$ 4. NC; NC   5. $\downarrow$; NC

C. 5.3%

D. 1. 3.7   2. $P_A = 0.41$ atm; $P_B = 1.2$ atm; $P_C = 1.3$ atm   3. 1.9

Worksheet C
A. 1. F   2. F   3. T   4. T   5. F
   6. F   7. F   8. F   9. F   10. T
   11. T   12. T   13. T   14. F   15. F

B. 1. a. $\uparrow$ b. NC   c. $\downarrow$
   2. a. $\downarrow$ b. $\downarrow$ c. NC   d. $\uparrow$, $\downarrow$, $\uparrow$

C. 1. 0.96 atm
   2. yes
   3. $P_{CO_2} = P_{H_2O} = 0.48$ atm; 17 g Na_2CO_3 ; 15 g $NaHCO_3$

D. 1. $P_A = 2.5$ atm; $P_B = 0.7$ atm; $P_C = 1.0$ atm
   2. $n_D = 0.88$ mol
   3. 307 K = 34°C

Acids and Bases

Worksheet A

A. Circle the correct answer(s).

1. A water solution is prepared by dissolving 0.20 mol of a weak acid, HB, in water to form one liter of solution. One can write the following relations about the species in solution when equilibrium is established.

   a. $[HB] = 0.20 - ([H^+] + [B^-])$ b. $[H^+] < 0.20$ M
   c. $[H^+] = [B^-]$ d. $[H^+] + [B^-] = 0.20$
   e. $[H^+] + [HB] \approx 0.20$ f. $pH = -\log (0.20)$

2. The pH of a solution of HCl that is 1.0×10^{-9} M at 25°C is
   a. 9.00.
   b. less than 7.00.
   c. 7.00.
   d. less than 9.00 but greater than 7.00.
   e. can't tell.

3. Consider a 0.100 M aqueous solution of HNO_3.
   a. $[H^+] = [NO_3^-] = [HNO_3]$
   b. The conductivity of the solution is the same as that of a 0.100 M aqueous solution of HF.
   c. The freezing point of the solution is lower than that of a 0.100 M aqueous solution of $HClO_4$.
   d. The pH of the solution is 1.00.
   e. The pH of the solution is higher than that of a 0.100 M aqueous solution of HI.
   f. The percent ionization is 100%.

4. K_a for $HC_2H_3O_2$ (acetic acid) is 1.8×10^{-5}. K_b for NH_3 (aq) is also 1.8×10^{-5}. When equal volumes of 0.10 M solutions of acetic acid and ammonia are mixed, the pH of the resulting solution
   a. is 7 because the strength of the two solutions is comparable.
   b. is less than 7, because NH_3 (aq) is a stronger base than $HC_2H_3O_2$ is an acid.
   c. is greater than 7, because $HC_2H_3O_2$ is a stronger acid than NH_3 (aq) is a base.
   d. cannot be determined without a pH meter or more information.
   e. is 1 because $K_a/K_b = 1$.

5. Consider a weak acid with a pK_a of 5.00. The % ionization of a 0.10 M solution is
   a. 10 times that of a 0.01 M solution.
   b. 1/10 that of a 0.01 M solution.
   c. the same as that of a 0.01 m solution.
   d. larger, but not 10 times that of a 0.01 M solution,
   e. smaller, but not one tenth that of a 0.01 M solution

B. Arrange the following aqueous solutions in order of increasing pH.
   1.   4.0 M NaCl   1.0 M Sr(OH)$_2$ 2.0 M HClO$_3$ 1.5 M KOH   0.2 M HCl
   2. 0.10 M solutions of
            NH$_4$CN   FeCl$_3$ LiClO$_4$ Ba(OH)$_2$ HNO$_3$

C. Calculate the pH of a solution prepared by mixing 2.500 g of NaNO$_2$ in enough water
    to make 500.0 mL of solution.

D. The figure below represents an aqueous solution of a weak acid at equilibrium. Each
    square represents a 0.01 mole of H$^+$, each circle 0.01 mole of B$^-$. Combinations of
    a square and a circle represent the weak acid molecule, HB. The box represents a
    liter. Water molecules and the H$^+$ and OH$^-$ formed from the ionization of water, while
    present, are not shown.

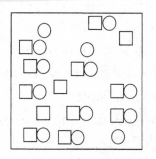

    Calculate: K$_a$, % ionization, and pH of the solution.

E. Calculate the pH of a solution made up by adding 0.100 g of NaOH to 250.0 mL of a
    0.200 M solution of Ba(OH)$_2$.

Acids and Bases

Worksheet B

A. Decide whether the statements are true or false. If the statement is false, rewrite it to make it true
   1. A 1.0 M aqueous solution of a strong acid will have a higher pH than a 1.0 M aqueous solution of a weak acid.
   2. Water acts as a Bronsted–Lowry base in the following equation.
$$NO_2^- \, (aq) + H_2O \rightleftharpoons OH^- \, (aq) + HNO_2 \, (aq)$$
   3. $Sr(C_2H_3O_2)_2$ forms an acidic solution in water.
   4. If the K_a for $H_2PO_4^-$ is 6.2×10^{-8}, then its K_b must be 1.6×10^{-7}.
   5. $CrCl_3$ forms an acidic solution in water.
   6. A 0.10 M solution of HB, a weak acid, has a lower percent ionization than a 0.010 M solution of the same weak acid, HB.
   7. $HClO_4$ is a weaker acid than $HClO_3$.
   8. It is possible for a weak acid to be more concentrated than a strong acid.
   9. When NaF is added to a solution of HF, the concentration of HF increases.
   10. A 1.0×10^{-8} M solution of HCl has a pH that is less than 7.

B. Answer the following questions.
   1. Consider 4 test tubes, each containing 100.0 mL of the following aqueous solutions.

A = 0.10 M HBr	B = 0.10 M HNO_2
C = 0.10 M NaOH	D = 0.10 M NH_3 (aq)

   Fill in the blanks with $<$, $>$, or $=$.
   a. pH of A _____ pH of B.
   b. pH of B _____ 1 .
   c. pH of C _____ pH of D.
   d. % ionization of A _____ % ionization of C.
   e. When 1.00 L of solution C is added to 1.00 L of solution D, a solution is obtained where the moles of OH^- _____ 0.05 .

2. Acetic acid, $HC_2H_3O_2$, is most often written as CH_3COOH, where the ionizable hydrogen is the H in the COOH group. The following series of acids are those where the H's in the CH_3 group are replaced by Cl.

acid	K_a
CH_3COOH	1.8×10^{-5}
$Cl–CH_2–COOH$	1.4×10^{-3}
$Cl–CH–COOH$	3.3×10^{-2}
$\quad\;\; |$	
$\quad\;\; Cl$	
$\quad\;\; Cl$	
$\quad\;\; |$	
$Cl–C–COOH$	2.0×10^{-1}
$\quad\;\; |$	
$\quad\;\; Cl$	

a. What trend in acid strength do you observe?
b. What can you say about the ionizability of the H in the COOH group and Cl substitution in the CH_3 group?
c. Suppose that you had 0.10 M aqueous solutions of each of the 4 acids. Which of the 4 would have the highest pH; the highest % ionization?
d. Compare the pH of a 0.10 M aqueous solution of HCl with that of a 0.10 M aqueous solution of CCl_3COOH (trichloroacetic acid), a common ingredient in solutions advertised as "wart removers".

C. Calcium hydroxide is only slightly soluble in water (0.50 g/L water). Assuming that 0.50 g of $Ca(OH)_2$ in 1.000 L of water results in 1.000 L of solution, what is the maximum pH one can obtain from a saturated solution of $Ca(OH)_2$?

D. What volume of HCl gas at 25°C and 1 atm should be bubbled into 375 mL of a 0.680 M aqueous solution of $HClO_4$ so that the pH of the resulting solution is 0? Assume that all the HCl dissolves and that the volume of the solution does not change with the addition of HCl.

E. What is the pH of a solution prepared by adding enough water to 15.00 g of $NaNO_2$ to make 275 mL of solution?

Acids and Bases

Worksheet C

A. Circle the correct answer(s).

1. When NaF is added to a solution of HF
   a. the pH increases.   b. $[H^+]$ increases.   c. K_b increases.
   d. [HF] increases.   e. K_a decreases.

2. Consider a 0.01 M aqueous solution of a weak acid HB.
   a. $[B^-]$ is always greater than [HB].
   b. Its conductivity equals that of 0.10 M HCl.
   c. Its freezing point is higher than that of 0.10 M HCl.
   d. The solution has a pH of 2.0 .
   e. $[B^-] = [H^+] = [HB]$

3. The ionization constant for H_2S is 1.0×10^{-7}, while that for HS^- is 1.3×10^{-13}. For 0.10 M aqueous solutions of Na_2S and NaHS,
   a. both solutions have a pH of 7.0.
   b. Na_2S is more basic than NaHS.
   c. both are acidic.
   d. K_b for HS^- is 7.7×10^{-2}.
   e. HS^- is an amphiprotic ion.

4. The following conditions increase the percent ionization of a weak acid, HB.
   a. Addition of a strong acid
   b. Addition of a salt, NaB
   c. An increase in temperature
   d. Diluting the acid with more water
   e. Nothing can change percent ionization. It is a constant.

5. Consider the reaction

   $$HF\,(aq) + NH_3\,(aq) \rightleftharpoons NH_4{}^+\,(aq) + F^-\,(aq) \qquad K = 1.2 \times 10^6$$

   The following can be said about this reaction.
   a. The conjugate base of HF is F^-.
   b. HF and NH_3 are a conjugate acid–base pair.
   c. When equilibrium is reached, the predominating species will be HF and NH_3.
   d. This is not an acid–base reaction because neither H^+ nor OH^- is present.
   e. Since K is large, there are no HF or NH_3 molecules in the solution after reaction is complete.

B. Using the following symbols:

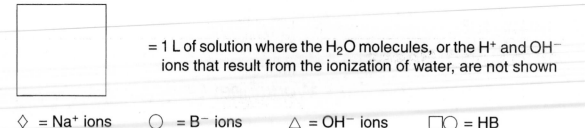

= 1 L of solution where the H_2O molecules, or the H^+ and OH^- ions that result from the ionization of water, are not shown

◇ = Na^+ ions   ○ = B^- ions   △ = OH^- ions   ▢○ = HB

1. Draw a picture to represent a solution in which 30% of B^- becomes HB according to the equation

$$B^- (aq) + H_2O \rightleftharpoons HB(aq) + OH^- (aq)$$

Do not represent the water molecules, and start with 0.10 moles of NaB. (Each square, triangle, etc. can represent 0.01 mole.)

2. Calculate K_b for the picture that you drew in (1).

3. Calculate the pH of the solution in the picture.

C. Diet Coke has a pH of about 3.0. Milk has a pH of about 7.0. How much greater is the H^+ ion concentration of Diet Coke when compared to milk?

D. Calculate the pH of a solution made up by adding 10.0 g NaOH and 15.0 g $Ba(OH)_2$ to enough water to make 0.500 L of solution.

E. What is the K_a of a weak acid if 0.0342 mole of weak acid are added to enough water to make 625 mL of solution that has a pH of 3.79?

Acids and Bases

Answers

Worksheet A

A. 1. b, c, e 　　　2. b 　　　3. d, f 　　　4. a 　　　5. e

B. 1. $0.2\,M\,HCl < 2.0\,M\,HClO_3 < 4.0\,M\,NaCl < 1.5\,M\,KOH < 1.0\,M\,Sr(OH)_2$
　2. $HNO_3 < FeCl_3 < LiClO_4 < NH_4CN < Ba(OH)_2$

C. 8.04

D. $K_a = 9.0 \times 10^{-3}$; % ionization = 23%; pH = 1.5

E. 12.07

Worksheet B

A. 1. False — lower pH
　2. False — Bronsted–Lowry acid
　3. False — basic
　4. False — K_b for its conjugate base HPO_4^{2-} is 1.6×10^{-7}
　5. True
　6. True
　7. False — $HClO_4$ is a stronger acid than $HClO_3$
　8. True
　9. True
　10. True

B. 1. a. < 　　　b. > 　　　c. > 　　　d. = 　　　e. >

　2. a. acid strength increases as the number of Cl in the CH_3 group increases.
　　b. The H in the COOH group becomes more ionizable as the number of Cl in the CH_3 group increases.
　　c. CH_3COOH — highest pH; CCl_3COOH — highest % ionization
　　d. pH of 0.10 M HCl = 1.00; pH of 0.10 M CCl_3COOH = 1.14

C. 12.13

D. 2.94 L

E. 8.56

Worksheet C

A. 1. a, d   2. c   3. b, e   4. d   5. a

B. 1.   2. $K_b = 0.0133$ 3. pH = 12.48

C. 1×10^4 times greater

D. 13.93

E. 4.82×10^{-7}

Equilibria in Acid–Base Solutions

Worksheet A

A. Consider the solutions below, each with a volume of 1.00 L. In the first blank, write **Y** if the solution is a buffer and **X** if it is not a buffer. In the next blank, write **A** if the solution is acidic, **B** if the solution is basic, and **N** if the solution is neutral. K_b for NH_3 (aq) is 1.8×10^{-5}.

1. _____ , _____ 0.10 mol HCl + 0.10 mol KOH

2. _____ , _____ 0.10 mol HCl + 0.30 mol KOH

3. _____ , _____ 0.10 mol HCl + 0.050 mol NH_3

4. _____ , _____ 0.10 mol HCl + 0.500 mol NH_3

5. _____ , _____ 0.10 mol HCl + 0.500 mol NH_4^+

6. _____ , _____ 0.10 mol HCl + 0.500 mol NH_3 + 0.300 mol NH_4^+

7. _____ , _____ 0.10 mol HCl + 0.050 mol NH_3 + 0.300 mol NH_4^+

B. Consider the titration curve below. The titrating agent (titrant) is the solution in the buret. Both solutions are 0.100 M. There are 50.0 mL of solution to be titrated.
1. Is the titrating agent an acid or a base?
2. Is the solution to be titrated (solution in the beaker) an acid or base? strong or weak?
3. What is the pH at the equivalence point?
4. What is the K_a of the solution being titrated?
5. What pH or pK_a range should the indicator for this titration have?

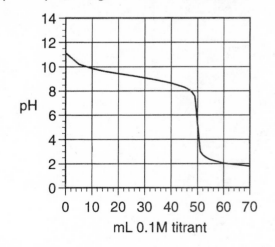

mL 0.1M titrant

C. Calculate the pH of a solution made up of 25.0 g of NH_4Cl and 10.0 g of NaOH in enough water to make 0.300 L of solution.

D. Consider the titration of NH_4Cl with KOH.
1. Calculate the pH before titration, at half-neutralization, and at the equivalence point when of 25.00 mL of 0.200 M NH_4Cl are titrated with 0.100 M KOH.

2. Draw a rough titration curve for a strong base-weak acid titration.

3. At what pH should a suitable indicator change for this titration?

Equilibria in Acid–Base Solutions

Worksheet B

A. Circle the correct answer(s).
  1. Consider the titration represented by the following equation:

$$H^+ (aq) + F^- (aq) \rightleftharpoons HF (aq)$$

  a. The titration could be the net ionic equation for the titration of NaF with H_2SO_4.
  b. The reaction could be considered to be that of a strong acid with a strong base.
  c. At the equivalence point, the resulting solution can act as a buffer.
  d. Diluting the solution obtained at half-neutralization with pure water does not change the pH of the solution.
  e. Phenolphthalein (pH range: 9–11) would be an effective indicator in this titration.

  2. Consider 0.100 M solutions of acetic acid, NaOH, and HCl. A buffer is formed when
  a. 20.00 mL of each solution are mixed.
  b. 20.00 mL of acetic acid and 30.00 mL of NaOH are mixed.
  c. 30.00 mL of acetic acid and 20.00 mL of NaOH are mixed.
  d. 20.00 mL of acetic acid and 20.00 mL of NaOH and 10.00 mL of HCl are mixed.
  e. acetic acid is titrated with NaOH to half-neutralization.

  3. In the titration of a weak base (e.g., NH_3) by a strong acid (e.g., HCl), the highest pH on the titration curve is
  a. the initial pH.   b. the pH at half-neutralization.
  c. the pH at the equivalence point.   d. the pH beyond the equivalence point.

B. A 0.100 M solution of a weak acid, HX, is known to be 15% ionized. The weak acid has a molar mass of 72 g/mol.
  1. What is K_a for the weak acid?
  2. What is the pH of the buffer prepared by adding 10.0 g of the sodium salt of the acid (NaX) to 100.0 mL of 0.250 M HX?

C. Fill the blanks with the appropriate symbol or word.
1. Two buffers are prepared in the following manner.

Buffer	[HX]	[X⁻]
A	0.50 M	0.20 M
B	0.25 M	0.20 M

Complete the statements using $<$, $>$, or $=$

a. pH of Buffer A _____ pH of Buffer B.

b. capacity to absorb acid: Buffer A _____ Buffer B.

c. capacity to absorb base: Buffer A _____ Buffer B.

d. $[H^+]$ of Buffer A _____ $[H^+]$ of Buffer B.

2. Consider this series of titration curves for 50.0 mL of 0.10 M solutions (A – E):

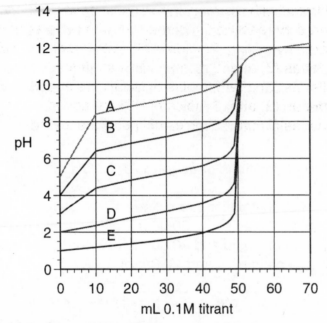

a. The titrant is (acid or base) _____ .

b. The titration curve that could represent the titration of a strong acid with a strong base is _____ .

c. The pH range for an appropriate indicator for titration C is _____ .

d. The titration curve that represents the titration of the weakest acid is _____ .

e. The K_a for solution C is approximately _____ .

D. Represent 1 mole of the following species with the the given symbols: H^+ – ▢ ,
    B^- – ◯ , OH^- – △ , HB – ▢◯   . A box represents a volume of 1L. Thus the picture
    below represents 2M HB.

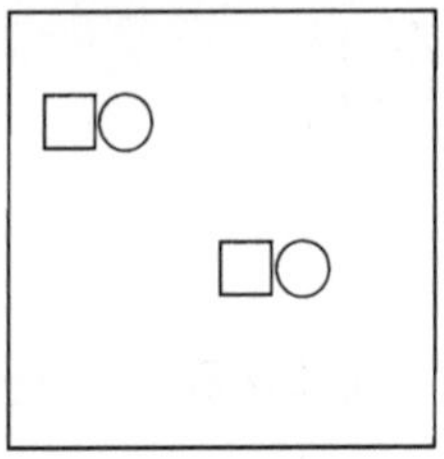

▢◯

▢◯

Water molecules and the few H^+ and OH^- ions from the ionization of water are not
represented.

Show pictorially the following situations:

1. A weak acid titrated with a strong base at half–neutralization.
2. A buffer prepared by reacting 5 moles of B^- with 2 moles of H^+.
3. A buffer prepared by mixing 4 moles of B^- with 3 moles of HB.
4. The same buffer as (3) after 2 moles of HCl are added.
5. The same buffer as (3) after 2 moles of NaOH are added.
6. The same buffer as (3) after 5 moles of HCl are added.
7. The same buffer as (3) after 5 moles of NaOH are added.

Equilibria in Acid–Base Solutions

Worksheet C

A. Consider the following titration curves. Titration curve A represents 50.00 mL of 0.100 M solution A titrated with 0.100 M of a titrant. Titration curve B represents 50.00 mL of 0.100 M solution B titrated with the same titrant. Answer the following questions.

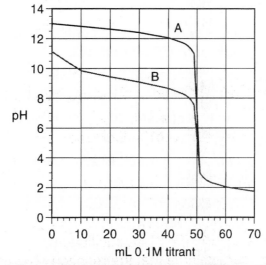

1. Describe A and B as either acid or base, strong or weak.
2. The titrant must be a(n) (acid or base).
3. Which of the two solutions has a higher % ionization?
4. How many mL are required for the half-neutralization of A?
5. What is an approximate K_a for B?
6. What is the pH at the equivalence point for the titration of B?
7. Can methyl red (pH range: 4–6) be used as a suitable indicator for both titrations?

B. Consider the solutions below. In the first blank, write **Y** if the solution is a buffer and **X** if it is not a buffer. In the next blank, write **A** if the solution is acidic, **B** if the solution is basic, and **N** if the solution is neutral.
K_a for HF (aq) is 6.9×10^{-4}.

1. _____ , _____ 25.00 mL of 0.100 M KOH titrated with HF to its equivalence point.

2. _____ , _____ One mol of KCl and 1 mol of HCl in enough water to make 1 liter of solution.

3. _____ , _____ One mol of $NaNO_2$ and 1 mol of NaF in enough water to make 1 liter of solution.

4. _____ , _____ 25.00 mL of HF titrated half-way to the equivalence point with NaOH.

5. _____ , _____ One liter of a solution containing 0.100 mol HF, 0.100 mol NaF, and 0.0200 mol HCl.

6. _____ , _____ One liter of a solution containing 0.100 mol HF and 0.200 mol NaOH.

7. _____ , _____ One liter of a solution containing 0.100 mol HF, 0.100 mol NaF, and 0.150 mol HCl.

8. _____ , _____ A solution of 0.100 M HF titrated to its equivalence point with KOH.

9. _____ , _____ Twenty five mL of 0.100 M HCl titrated with 10.00 mL of 0.100 M NaOH.

10. _____ , _____ One liter of a solution containing 0.200 mol NaF and 0.100 mol HCl.

C. You are asked to prepare one liter of a buffer with a pH of 5.4. Your lab only has 1.000 M solutions of acids and bases. The stockroom which has solid salts is closed. Describe how you would prepare the buffer using acetic acid ($HC_2H_3O_2$) and NaOH. [*Hint:* Find $[C_2H_3O_2^-]/[HC_2H_3O_2]$ first.]

D. Consider the titration of 100.0 mL of 0.100 M HCl with 0.200 M NaOH.
   1. What is the pH after the addition of the following amounts of NaOH in the titration?
      a. 0.00 mL
      b. 10.00 mL
      c. 49.00 mL
      d. 50.00 mL
      e. 51.00 mL
      f. 60.00 mL
   2. Sketch the titration curve for this experiment using the data obtained in (1).

Equilibria in Acid–Base Solutions

Answers

Worksheet A

A. 1. X, N   2. X, B   3. X, A   4. Y, B
   5. X, A   6. Y, B   7. X, A

B. 1. acid   2. weak base   3. ≈ 5 4. $\approx 1 \times 10^{-9}$
   5. $4 - 6$

C. 9.31

D. 1. 0–time: 4.98; half-neutralization: 9.25; equivalence point: 11.04
   2.

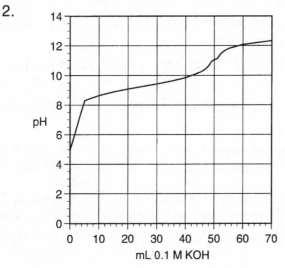

mL 0.1 M KOH

   3. $10 - 12$

Worksheet B

A. 1. a, d   2. c, d, e   3. a

B. 1. 2.6×10^{-3} 2. 3.23

C. 1. a. $<$ b. $=$ c. $>$ d. $>$
   2. a. base   b. E   c. anything that changes pH from 7 to 10
      d. A   e. $\approx 1 \times 10^{-5}$

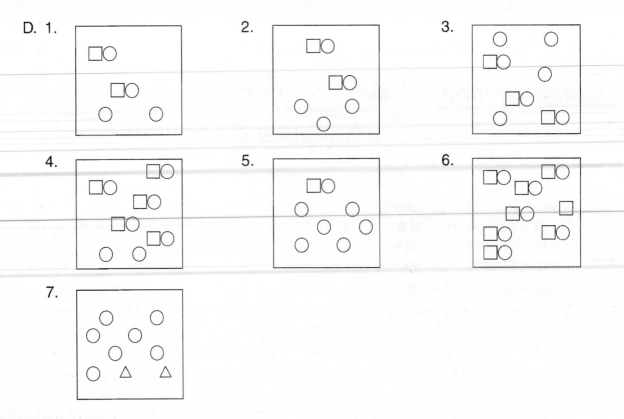

D. 1.   2.   3.

4.   5.   6.

7.

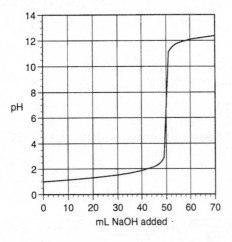

Worksheet C

A. 1. A - strong base; B - weak base   2. acid   3. A
   4. 25 mL   5. $\approx 1 \times 10^{-9}$ 6. ≈ 5 7. yes

B. 1. X, B   2. X, A   3. X, B   4. Y, A   5. Y, A
   6. X, B   7. X, A   8. X, B   9. X, A   10. Y, A

C. Answers may vary. $n_{Ac^-}/n_{HAc} = 4.5$ Most common answer: 0.55 L HAc and 0.45 L NaOH

D. 1. a. 1.00   b. 1.14   c. 2.87   d. 7.00   e. 11.12   f. 12.10
   2.

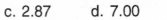

Complex Ions

Worksheet A

A. Answer the following questions.
1. Write the formula of a complex ion made up of zinc(II) and the following ligands: two thiocyanate, one ethylenediamine, and one oxalate.
2. Write the formula for the salt (sodium or chloride, whichever is possible) of the complex ion described in (1).
3. For which of the following transition metal ions is it not possible to have both low spin and high spin octahedral complexes?
$$Ni^{2+},\ Fe^{3+},\ Fe^{2+},\ Co^{3+},\ Mn^{2+}$$
4. How many unpaired electrons are there in the high spin and low spin complexes of octahedra with Ru^{2+} as the central atom?
5. Complexes that have empty low–energy d orbitals or those that have occupied higher d orbitals generally are labile (ligands are exchanged rapidly). The others are expected to be inert. On the basis of the electron configuration of the central atom, which of the following complexes are expected to be labile? Why?

   a. $Cr(NH_3)_6^{3+}$ b. $Co(H_2O)_6^{2+}$ (high spin)

   c. $Ti(H_2O)_6^{3+}$ d. $Fe(CN)_6^{3-}$ (low spin)

B. Write all possible geometric structures for
1. $[Ag(NH_3)H_2O]^+$ 2. $[Cu(H_2O)_3Cl]^+$ 3. $[Co(NH_3)_4Cl_2]^+$
4. $[Cr(en)Cl_4]^{2-}$ 5. $[Co(en)Cl_2I_2]^-$

C. For $Cd(NH_3)_4^{2+}$ at a certain temperature, $K_f = 2.8 \times 10^7$.
1. Write the K_f expression for the formation of this ion.
2. What is $[NH_3]$ if $[Cd^{2+}] = [Cd(NH_3)_4^{2+}]$?
3. At what pH does the concentration of NH_3 calculated in (2) occur? (K_b for NH_3 is 1.8×10^{-5}.)

Complex Ions

Worksheet B

A. Circle the true statement(s).

1. Consider three complexes of Ag^+ and their formation constants, K_f.

Complex ion	K_f
$Ag(NH_3)_2^+$	1.6×10^7
$Ag(CN)_2^-$	5.6×10^{18}
$AgBr_2^-$	1.3×10^7

   a. $Ag(NH_3)_2^+$ is more stable than $Ag(CN)_2^-$.

   b. Adding a strong acid (HNO_3) to a solution that is 0.010 M in $Ag(NH_3)_2^+$ will tend to dissociate the complex ion into Ag^+ and NH_4^+.

   c. Adding a strong acid (HNO_3) to a solution that is 0.010 M in $AgBr_2^-$ will tend to dissociate the complex ion into Ag^+ and Br^-.

   d. To dissolve AgI, one can add either NaCN or HCN as a source of the cyanide complexing ligand. Fewer moles of NaCN would be required.

   e. Solution A is 0.10 M in Br^- and contains the complex ion $AgBr_2^-$. Solution B is 0.10 M in CN^- and contains the complex ion $Ag(CN)_2^-$. Solution B will have more particles of complex ion per particle of Ag^+ than solution A.

2. Consider the coordination number of a complex ion.

   a. Most metal ions exhibit only a single characteristic coordination number.

   b. The coordination number is equal to the number of ligands bonded to the metal atom.

   c. The coordination number is determined solely by the tendency to surround the metal atom with the same number of electrons as one of the rare gases.

   d. There are no coordination numbers above six.

   e. For most complexes, the coordination number is equal to the number of bonds formed with the metal atom.

3. A bidentate ligand

   a. has bonds formed to two metal atoms.

   b. has a charge of either +2 or –2.

   c. forms complex ions that have a charge of +2 or –2.

   d. has two atoms that can donate an electron pair.

   e. has a coordination number of 2.

   f. forms two bonds with the central metal atom.

4. A low spin complex
   a. is formed by weak field ligands.
   b. has all d electrons unpaired.
   c. has a large crystal field splitting energy.
   d. is always paramagnetic.
   e. concentrates the electrons in the lower energy orbitals.

5. Complexes of Zn^{2+}, a d^{10} ion, are all colorless. The most likely explanation for this is that
   a. Zn^{2+} is paramagnetic.
   b. Zn^{2+} complexes exhibit d–orbital splitting such that they absorb all wavelengths in the visible region.
   c. Zn^{2+} cannot absorb visible light even though there is splitting of d–orbitals.
   d. Zn^{2+} absorbs in the yellow region which is colorless to the naked eye.

B. Consider $[Cr(Br)_4Cl_2]^{4-}$.
   1. Write the electron configuration of the chromium ion.

   2. Write the orbital structure of the chromium ion in both low spin and high spin octahedral complexes.

   3. In which complex (high spin or low spin) is the ion paramagnetic?

   4. Draw all possible isomers for the complex ion.

   5. Write the formula for the aluminum salt of this ion.

   6. At some temperature, K_f for this ion is 3×10^{15}. What is the ratio of the complex ion to the metal ion in a solution that is 0.10 M in both Br^- and Cl^-?

Complex Ions

Worksheet C

A. Decide whether the statements below are true or false. If the statement is false, rewrite it to make it true.

1. A metal atom, coordination number 6, which can form both low and high spin complexes will have fewer unpaired electrons in the low spin complex.

2. A complex of the form MCl_2Br_2 (M = metal) which exists in two isomeric forms must be tetrahedral.

3. If the configuration of the metal ion is d^0, d^1, d^2, or d^3, the complex will have the same high spin and low spin orbital diagram.

4. The charge on the central metal atom in $Pt(NH_3)_4^{2+}$ is +2.

5. Typical bidentate ligands include ethylenediamine (en), oxalate (ox), and EDTA (ethylenediamine tetracetate ion).

6. The octahedral complexes of Ni^{3+} are diamagnetic for both high spin and low spin complexes.

7. For the complex ion $[Co(en)(OH)_2(H_2O)_2]^+$, the charge on the central atom is +1.

8. For the same complex ion described in (7), the high spin octahedral complex has no unpaired electrons.

9. There are three different geometric isomers possible for the complex ion described in (7).

10. The coordination number for the complex ion described in (7) is 5.

11. Ethylenediamine (en) is a bidentate ligand.

12. Octahedral complexes can be low spin or high spin.

13. Strong field ligands yield high spin complexes whereas weak field ligands yield low spin complexes.

14. Octahedral d^{10} complexes are colored due to transitions between two types of orbitals.

15. Paramagnetic compounds are attracted by a magnetic field.

16. The central metal atom in a complex donates a pair of electrons to the ligand to form a covalent bond.

17. No metal in period 3 of the periodic table can be a central atom because period 3 has no transition metals.

18. The d_{z^2} and $d_{x^2-y^2}$ orbitals are split from the d_{xy}, d_{yz}, and d_{xz} orbitals in the octahedral field.

19. In complex ions, the ligands act as Lewis bases.

20. The d_{z^2} and $d_{x^2-y^2}$ orbitals have a higher energy than the d_{xy}, d_{yz}, and d_{xz} orbitals.

B. Answer the following questions.

1. What is the formula for the sulfate salt of $Ni(en)(CN)_2(H_2O)_2^+$?

2. Draw all the isomers of $Pt(ox)(en)Cl_2$.

3. The origin of color in coordination complexes is attributed to the visible light absorbed when a d electron is excited from the low energy d_{xy}, d_{yz}, or d_{xz} orbital into a vacancy in the higher energy $d_{x^2-y^2}$ and d_{z^2} orbitals. If such a transition is to take place, the d subshell of the metal must be partially filled; that is, it can be neither completely empty nor completely filled. Knowing this, predict which of the following complexes will be colored.

$$Cd(NH_3)_4^{2+}, \ Fe(H_2O)_6^{3+}, \ Sc(H_2O)_6^{3+}, \ Co(NH_3)_6^{3+}$$

C. The K_f at a certain temperature for $Ag(OH)_2^-$ is 3×10^7. At what pH would $[Ag(OH)_2^-]$ $= 10 \ [Ag^+]$?

Complex Ions

Answers

Worksheet A

A. 1. $Zn(SCN)_2(en)(ox)^{2-}$ 2. $Na_2[Zn(SCN)_2(en)(ox)]$

  3. Ni^{2+} 4. low spin: 0; high spin: 4

  5. a. inert — no empty low energy d orbitals; no occupied high energy d orbitals

      b. labile — high energy d orbitals are occupied

      c. labile — two low energy d orbitals are empty

      d. inert — no empty low energy d orbitals; no occupied high energy d orbitals

B. 1. $[H_2O - Ag - NH_3]^+$ 2.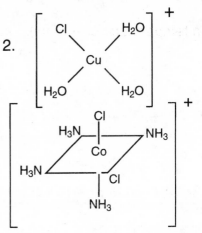

3.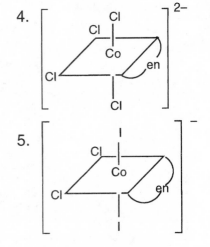

4.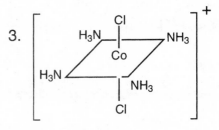

5.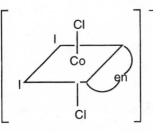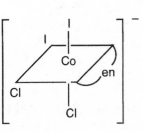

C. 1. $K_f = \dfrac{[Cd(NH_3)_4^{2+}]}{[Cd^{2+}][NH_3]^4}$ 2. $[NH_3] = 0.014$ 3. 10.70

Worksheet B

A. 1. b, d, e   2. e   3. d, f   4. c, e   5. c

B. 1. $1s^2\ 2s^2\ 2p^6\ 3s^2\ 3p^6\ 3d^4$

2.   low spin   high spin

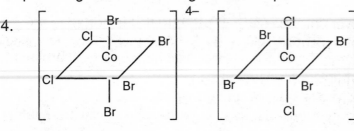

3. paramagnetic for both high and low spin

4.

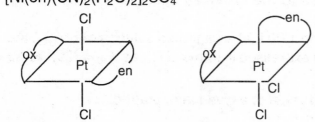

5. $Al_4[Cr(Br)_4(Cl)_2]_3$ 6. 3×10^9

Worksheet C

A. 1. True   2. False — must be square planar
   3. True   4. True
   5. False — polydentate ligands   6. False — paramagnetic
   7. False — +3   8. False — low spin
   9. True   10. False — 6
   11. True   12. True
   13. False — Strong field ligands yield low spin complexes. Weak field ligands yield high spin complexes.
   14. False — colorless due to no transitions
   15. True   16. False — accepts a pair...
   17. False — Al can be a central atom   18. True
   19. True   20. True

B. 1. $[Ni(en)(CN)_2(H_2O)_2]_2SO_4$

2.

3. Only $Co(NH_3)_6^{3+}$ and $Fe(H_2O)_6^{3+}$.

C. 10.8

Precipitation Equilibria

Worksheet A

A. Circle the correct answer(s).

1. When 0.10 mol of solid silver chloride (K_{sp} = 1.8×10^{-10}) is added to 1.0 L of a clear, <u>saturated</u> solution of Ag_2CrO_4 (K_{sp} = 2.4×10^{-12}), which of the following result?
   a. $[Cl^-]$ = 0.10 M after equilibrium is established.
   b. The concentration of Ag^+ will change.
   c. The concentration of CrO_4^{2-} decreases.
   d. Some Ag_2CrO_4 will precipitate.
   e. All the silver chloride will dissolve.

2. The best explanation for the fact that ZnS dissolves in dilute HCl is that
   a. Zn^{2+} is amphoteric.
   b. $[S^{2-}]$ decreases because H_2S (g) is formed.
   c. $[S^{2-}]$ decreases because S (s) is formed.
   d. $[Zn^{2+}]$ decreases because $ZnCl_4^{2-}$ is formed.
   e. The K_{sp} of $ZnCl_2$ is less than the K_{sp} of ZnS.

3. Consider dissolving Sb_2S_3 in water.
$$Sb_2S_3 \text{ (s)} \rightleftharpoons 2\,Sb^{3+} \text{ (aq)} + 3\,S^{2-} \text{ (aq)}$$
   a. The molar solubility, s, of Sb_2S_3 can be expressed in terms of K_{sp} as $K_{sp} = 36\,s^5$.
   b. When sulfur powder, S (s), is added to the saturated solution, neither $[Sb^{3+}]$ nor $[S^{2-}]$ changes.
   c. Heating the solution increases the molar solubility of Sb_2S_3.
   d. When $Sb(NO_3)_3$ is added to the saturated solution, the K_{sp} for Sb_2S_3 increases.
   e. If 0.10 M Na_2S is added to 1.00 L of a saturated solution of Sb_2S_3, the molar solubility of Sb_2S_3 can be expressed in terms of K_{sp} as $K_{sp} = (2s)^2 (0.10+3s)^3$.

4. For an insoluble metallic salt,
   a. K_{sp} has a fixed numerical value at a given temperature.
   b. K_{sp} is always smaller than 1.
   c. adding a common ion does not change the concentration of the ions at equilibrium.
   d. K_{sp} cannot be determined.
   e. The unit for K_{sp} depends on the formula of the metallic salt.

5. Consider $PbBr_2$. Its molar solubility at 25°C is 1.2×10^{-2} M.
   a. A saturated solution has $[Pb^{2+}] = 1.2 \times 10^{-2}$ M and $[Br^-] = (1.2 \times 10^{-2})^2$ M.
   b. K_{sp} for $PbBr_2$ is 6.9×10^{-6}.
   c. Since K_{sp} increases when temperature is increased, the formation of $PbBr_2$ from its ions

   $$Pb^{2+} (aq) + 2\,Br^- (aq) \rightleftharpoons PbBr_2 (s)$$

   must be an exothermic reaction.
   d. Dissolving 6.0 g (0.016 mol) of $PbBr_2$ in water to make 2.0 L of solution results in an unsaturated solution of $PbBr_2$.
   e. When lead shot, Pb (s), is added to a saturated solution of $PbBr_2$, a precipitate of $PbBr_2$ can be expected because of the common ion effect.

B. Seventy mL of 0.0114 M $Sr(NO_3)_2$ are added to 30.00 mL of 0.0267 M K_2CrO_4. The resulting solution has a volume of 100.0 mL.

   1. Will $SrCrO_4$ ($K_{sp} = 3.6 \times 10^{-5}$) precipitate?

   2. Calculate $[CrO_4^{2-}]$ and $[K^+]$ after equilibrium is established.

   3. If more water is added to the solution in equilibrium to make its total volume 200.0 mL, will $SrCrO_4$ precipitate?

C. Consider CdS ($K_{sp} = 1.0 \times 10^{-29}$). Ammonia is added to form $Cd(NH_3)_4^{2+}$ ($K_f = 2.8 \times 10^7$).

   1. Calculate $[NH_3]$ if $[Cd^{2+}] = [Cd(NH_3)_4^{2+}]$.

   2. K_b for NH_3 is 1.8×10^{-5}. At what pH is the concentration of ammonia such that the condition in (1) holds?

   3. What is the molar solubility of CdS in 2.0 M NH_3 (aq)?

   4. Compare the molar solubility of CdS in pure water and in 2.0 M NH_3 (aq).

   5. Is NH_3 (aq) a "good" solvent for CdS?

Precipitation Equilibria

Worksheet B

A. Decide whether the following statements are true or false. If the statement is false, rewrite it to make it true.

1. In a saturated solution of $PbCl_2$ where $s = [Pb^{2+}]$, $[Cl^-] = 2s^2$.

2. The solubility, s, of $ScCl_3$ can be expressed in terms of K_{sp} as $9s^4$.

3. When solid $SrSO_4$ is added to a saturated solution of $SrSO_4$, the concentrations of Sr^{2+} and SO_4^{2-} do not change.

4. Heating always increases the molar solubility of a salt.

5. The K_{sp} for $BaCrO_4$ is 1.2×10^{-10} while that of $PbCrO_4$ is 2×10^{-14}. When CrO_4^{2-} ions are added to an aqueous solution of 0.10 M $Ba(NO_3)_2$ and 0.10 M $Pb(NO_3)_2$, $PbCrO_4$ will precipitate first.

6. It is possible to determine K_{sp} given a graph of the solubility of a salt in 100 g of water at different temperatures.

7. For the reaction
$$MX(s) + 4NH_3(aq) \rightarrow M(NH_3)_4^{2+}(aq) + X^{2-}(aq)$$
the equilibrium constant K is $\dfrac{K_{sp}}{K_f}$.

8. An amphiprotic species can dissolve in both acid and base.

9. When a solution is supersaturated, P is greater than K_{sp}, but no precipitation occurs.

10. A process used to separate solutes in a solution based on their solubility in the solvent is called fractional crystallization.

B. The compound $M(OH)_2$ ($\mathcal{M}$ = 100.0 g/mol) has the following solubility data:

T (°C)	mg $M(OH)_2$/100 g H_2O
10	10.0
15	22.0
20	34.0
30	56.0

1. Graph the solubility in pure water. Assume that the volume of the solution is the same as the mass of the water, i.e., 100 g H_2O = 100 mL of solution.

2. Calculate K_{sp} at the temperatures for which the solubility is given.

3. What is the solubility in mg $M(OH)_2$/100 g 0.10 M NaOH at the different temperatures? Assume that 100 g of 0.10 M NaOH is equal to 100 mL of a saturated solution of $M(OH)_2$ in 0.1 M NaOH.

4. Is the dissolving of $M(OH)_2$ in water an exothermic process?

C. Sixty milligrams of $SrSO_4$ are added to enough water to make 250.0 mL of solution. Will it all dissolve? How much remains undissolved if your answer is no?

D. Consider AgCl (K_{sp} = 1.8×10^{-10}) and $Ag(NH_3)_2^+$ (K_f = 1.7×10^7).
1. A solution has $[Ag^+]$ = 0.010 M from $AgNO_3$ and $[Cl^-]$ = 0.010 M from NaCl. Will AgCl precipitate?

2. A 1.00 M aqueous solution of NH_3 is added to a solution of 0.010 M $AgNO_3$. The following reaction occurs:

$$Ag^+ \text{ (aq)} + 2\,NH_3 \text{ (aq)} \rightleftharpoons Ag(NH_3)_2^+ \text{ (aq)}$$

a. What is $[NH_3]$ after it has reacted with Ag^+?

b. What is $[Ag^+]$ when equilibrium is established?

c. If 0.010 M NaCl is added after equilibrium is established, will AgCl precipitate?

Precipitation Equilibria

Worksheet C

A. Answer the following questions.

1. Consider the insoluble salts JQ, K_2R, L_2S_3, MT_2, and NU_3. They are formed from the metal ions, J^+, K^+, L^{3+}, M^{2+}, N^{3+}, and the nonmetal ions Q^-, R^{2-}, S^{2-}, T^-, and U^-. All the salts have the same K_{sp} at 25°C.
   a. Which salt has the highest molar solubility?

   b. Does the salt with the highest molar solubility have the highest solubility in g salt/100 g water?

   c. Can the solubility of each salt in g/100g water be determined from the information given? If yes, calculate the solubility of each salt in g/100g water. If no, why not?

2. Consider a beaker that has $M(OH)_2$ (s) and a saturated solution of $M(OH)_2$ $(K_{sp} = 1.0 \times 10^{-8})$.

   $$M(OH)_2 \text{ (s)} \rightleftharpoons M^{2+} \text{ (aq)} + 2OH^- \text{ (aq)} \qquad \Delta H > 0$$

   and $K_{sp} = 1.0 \times 10^{-8}$. What is the effect of the following on $[M^{2+}]$, $[OH^-]$, the mass of $M(OH)_2$ and K_{sp}? Write ↑ if there is an increase, ↓ if there is a decrease, and = if it remains the same.

	$[M^{2+}]$	$[OH^-]$	mass $M(OH)_2$	K_{sp}
a. add 0.10 M MX_2	——	——	——	——
b. increase T	——	——	——	——
c. add H_2O (ℓ)	——	——	——	——

3. What is likely to happen when fifty mL of 0.050 M $BaCl_2$ is added to fifty mL of 0.10 M K_2SO_4?
   a. KCl will precipitate.
   b. $[Ba^{2+}]_o = 0.025$ M after the two solutions are mixed (assume additive volumes) but before reaction occurs.
   c. If $[Ba^{2+}]_o \times [SO_4^{2-}]_o > [Ba^{2+}]_{eq} \times [SO_4^{2-}]_{eq}$, then precipitation will occur.
   d. $[Cl^-]_{eq} = [K^+]_{eq} = 5.0 \times 10^{-2}$ M

4. MX_2 can be represented as ⟨◯▢◯⟩ where circles represent X^- and squares represent M^{2+}. The molar solubility of MX_2 is 2 M. Fill in the empty box on the right to represent what happens when enough water is added to the box on the left to make 1 L of solution. (Each ◯▢◯ represents one mole.)

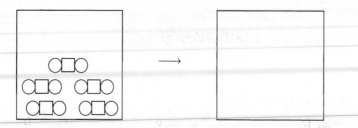

5. Write the reaction between $Zn(OH)_2$ and H^+. Write also the reaction between $Zn(OH)_2$ and OH^- to illustrate the amphoteric nature of $Zn(OH)_2$.

B. What is $[Al^{3+}]$ in a saturated solution of $Al(OH)_3$ ($K_{sp} = 2 \times 10^{-31}$) when the pH of the solution is 9.0?

C. Cadmium oxalate, CdC_2O_4, is relatively insoluble with a K_{sp} of 4×10^{-8}. Like Zn (directly above it in the periodic table) it forms the tetraammine complex $Cd(NH_3)_4^{2+}$ ($K_f = 5 \times 10^8$).
1. Calculate the molar solubility of CdC_2O_4 in water. Ignore the possible hydrolysis of $C_2O_4^{2-}$ and other competing reactions.

2. Calculate the molar solubility of CdC_2O_4 in a solution in which $[NH_3]_{eq}$ is 1.0 M. Assume that $Cd(NH_3)_4^{2+}$ is the only complex formed in a significant amount.

D. A saturated solution of $Mg(OH)_2$ is obtained by adding 0.145 g of $MgCl_2$ to 1.00 L of NH_3 (aq) with pH 9.80. What is the K_{sp} of $Mg(OH)_2$? Assume no volume changes.

Precipitation Equilibria

Answers

Worksheet A

A. 1. b, c, d 　　　　 2. b 　　　　 3. b, e 　　　　 4. a, b 　　　　 5. b, c, d

B. 1. yes
   2. $[CrO_4^{2-}] = 6.0 \times 10^{-3}\,M$; $[K^+] = 1.60 \times 10^{-2}\,M$
   3. no

C. 1. 0.014 M
   2. 10.70
   3. $6.7 \times 10^{-11}\,M$
   4. Molar solubility in water is $3.2 \times 10^{-15}\,M$ which is less than $6.7 \times 10^{-11}\,M$, the molar solubility in 2.0 M NH_3 (aq).
   5. 2.0 M NH_3 (aq) will dissolve more CdS than water, but it is not a particularly effective solvent.

Worksheet B

A. 1. False — $4s^2$ 　　　　　　　 2. False — $27\,s^4$
   3. True 　　　　　　　　　　　　 4. False — where $\Delta H_{solution} > 0$
   5. True 　　　　　　　　　　　　 6. True
   7. False — $K_{sp} \times K_f$ 　　　　 8. False — amphoteric
   9. True 　　　　　　　　　　　 10. False — fractional precipitation

B. 1.

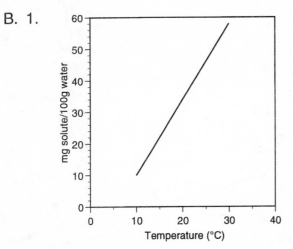

2.

T (°C)	K_{sp}
10	4.00×10^{-9}
15	4.26×10^{-8}
20	1.57×10^{-7}
30	7.02×10^{-7}

B. 3.

T (°C)	mg/100 g 0.10 M NaOH
10	0.00400
15	0.0426
20	0.157
30	0.702

    4. No

C. No;   0.0332 g

D. 1. yes

    2. a. $[NH_3] = 0.98$ M   b. $[Ag^+] = 6.1 \times 10^{-10}$ c. no

Worksheet C

A. 1. a. L_2S_3 b. not necessarily   c. No, $\mathcal{M}$ not given

    2.

	$[M^{2+}]$	$[OH^-]$	mass $M(OH)_2$	K_{sp}
a.	=	=	↑	=
b.	↑	↑	↓	↑
c.	=	=	↓	=

    3. b, c

    4.

    5. $Zn(OH)_2 \text{ (s)} + 2H^+ \text{ (aq)} \rightarrow Zn^{2+} \text{ (aq)} + 2H_2O$

       $Zn(OH)_2 \text{ (s)} + 2OH^- \text{ (aq)} \rightarrow Zn(OH)_4^{2-} \text{ (aq)}$

B. 2×10^{-16} M

C. 1. 2×10^{-4} M   2. 4 M

D. 6.1×10^{-12}

Spontaneity of Reaction

Worksheet A

A. Circle the correct answer(s).
1. Which of the following processes would you expect to be spontaneous?
   a. Dissolving NaCl in water at 25°C.
   b. Vaporizing water at 100°C.
   c. Melting candle wax at 200°C.
   d. Dry ice, CO_2 (s), subliming at 25°C.
   e. Making sugar ($C_{12}H_{22}O_{11}$) from water and carbon at 25°C.

2. A reaction for which enthalpy and entropy have the same sign
   a. is spontaneous at high T.
   b. is spontaneous at low T.
   c. can never be spontaneous.
   d. is spontaneous if the temperature changes in the right direction,
   e. is always spontaneous.

3. For a reaction, $\Delta G° = -70.0$ kJ at 298 K and + 10.0 kJ at 400 K. For this reaction
   a. $\Delta S° > 0$, $\Delta H° > 0$ b. $\Delta S° > 0$, $\Delta H° < 0$
   c. $\Delta S° < 0$, $\Delta H° > 0$ d. $\Delta S° < 0$, $\Delta H° < 0$

4. The following reaction

$$CH_4 (g) + 2 H_2S (g) \rightarrow CS_2 (g) + 4 H_2 (g)$$

   is endothermic. This reaction can be predicted to be spontaneous
   a. at low T only.   b. at high T only.   c. at all T.   d. at no T.

5. Which statements are true about entropy?
   a. As is true with enthalpies, absolute entropy values cannot be determined.
   b. At 0 K, even the movement of electrons ceases.
   c. Entropy always increases during a phase change.
   d. At 0 K, a perfect crystal has S = 0.
   e. Entropy usually increases when a molecule is broken into two small molecules.

B. Consider the graph below.

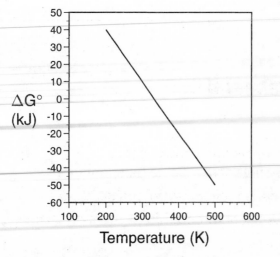

Temperature (K)

1. Describe the relationship between the spontaneity of the process and temperature.

2. Is the reaction exothermic?

3. Is $\Delta S° > 0$?

4. At what temperature is the reaction at standard conditions likely to be at equilibrium?

5. What is K for the reaction at 27°C?

C. At what temperature will acetone boil if $\Delta H_{vap} = 30.3$ kJ/mol, and $\Delta S_{vap} = 92.0$ J/mol-K?

D. Consider the reaction

$$Al_2O_3 \text{ (s)} + 2 \text{ Fe (s)} \rightarrow 2 \text{ Al (s)} + Fe_2O_3 \text{ (s)}$$

1. Is the reaction spontaneous at 100°C?

2. Is the spontaneity of the reaction temperature dependent? If so, at what temperature will the reaction be spontaneous?

3. Does the pressure have an effect on the spontaneity of this reaction?

Spontaneity of Reaction

Worksheet B

A. Circle the true statements. If the statement is false, make it true.

 1. Exothermic reactions are always spontaneous.

 2. Raising the temperature always increases the entropy of a solid.

 3. As T approaches 0 K, ΔH and ΔG become equal for a reaction involving only solids.

 4. At 25°C (standard temperature), a pure crystal of LiCl has $\Delta S° = 0$.

 5. $\Delta S°$ always increases during a phase change.

 6. As in the case with ΔH, ΔS depends on the initial and final state of a system, not on one particular pathway that was followed.

 7. $\Delta S°$ decreases during sublimation.

 8. Reactions where $\Delta G°$ are less than 0 always proceed spontaneously and rapidly.

 9. When $\Delta G_f° > 0$ for a compound, it means that the compound tends to be unstable and decomposes.

 10. $\Delta G°$ is not a state property and therefore is path dependent.

 11. A combustion reaction producing more moles of gas than are consumed will have a ΔG more negative than ΔH.

 12. For a spontaneous change, $\Delta G°_{total}$ must be negative for the overall equation between two coupled reactions.

 13. When a solute is dissolved in water, ΔS for the process is expected to be positive.

 14. Both heat capacity (C) and entropy (S) have the units J/K-mol and hence are the same thermodynamic property.

 15. A chemical reaction in which entropy decreases will never be spontaneous.

B. The relationship between $\Delta G°$ and T is linear. Draw the graph of a reaction with the properties described below. You need only label the point at which $\Delta G° = 0$.

 1. The reaction is exothermic.

 2. $\Delta n_g < 0$ (Δn_g = moles of gas products – moles of gas reactants)

 3. At 300 K, the system is at equilibrium and K = 1.

C. Acetic acid, CH_3COOH, freezes at 16.6°C. Its heat of fusion is 69.0 J/g. What is the change in entropy, ΔS, when one mol of acetic acid freezes?

D. At 25.0°C, K_{sp} for silver chloride is 1.782×10^{-10}. At 35.0°C, its K_{sp} is 4.159×10^{-10}. Calculate $\Delta H°$ and $\Delta S°$ for the reaction

$$Ag^+ (aq) + Cl^- (aq) \rightleftharpoons AgCl (s)$$

(*Hint* : Recall which thermodynamic properties are not temperature dependent.)

Spontaneity of Reaction

Worksheet C

A. Circle the correct answer(s).
1. Consider the following exothermic reactions.

$$(1)\ CH_4\,(g)\ +\ 2\,O_2\,(g)\ \rightarrow\ CO_2\,(g)\ +\ 2\,H_2O\,(g)$$
$$(2)\ C_2H_5OH\,(\ell)\ +\ 3\,O_2\,(g)\ \rightarrow\ 2\,CO_2\,(g)\ +\ 3\,H_2O\,(g)$$
$$(3)\ 2\,Al\,(s)\ +\ \tfrac{3}{2}\,O_2\,(g)\ \rightarrow\ Al_2O_3\,(s)$$

   a. ΔS for reaction (1) is zero because $\Delta n_g = 0$.
   b. ΔS for (2) is most probably positive.
   c. Equation (3) has a negative ΔS.
   d. Equation (2) is certainly spontaneous.
   e. The equations can be ranked (3) < (1) < (2) in terms of increasing ΔS.

2. Which statements about free energy are true?
   a. A combustion reaction producing heat and more moles of gas than are consumed will have a ΔG more negative than ΔH.
   b. Reactions where $\Delta G < 0$ will occur spontaneously and rapidly.
   c. When $\Delta G = 1$, the reaction is at equilibrium.
   d. Any form of each pure element is assigned ΔG_f°.
   e. When $Q = 1$, $\Delta G = \Delta G^\circ$

3. For the reaction

$$C_{(diamond)}\ \rightarrow\ C_{(graphite)} \qquad \Delta G^\circ < 0$$

   However, at room temperature and standard pressure (P = 1 atm), one does not observe diamond turning into graphite. An explanation for this is that
   a. the reaction is at equilibrium.

$$C_{(diamond)}\ \rightleftharpoons\ C_{(graphite)}$$

   Since the rate of the forward reaction is the same as the rate of the reverse reaction, no change is visible.
   b. diamond is thermodynamically stable.
   c. spontaneity does not imply rate, i.e., the reaction is occurring at such a slow rate that the change is not noticeable.
   d. diamond does become graphite but the rate at which graphite becomes diamond is much faster.
   e. allotropes do not follow the rules of free energy.

4. A solid and liquid are mixed. Observations include the following:
   — The solid disappears completely.
   — The beaker feels warmer to the touch after the compounds are mixed.
   — A gas is evolved.

   One can state with a high degree of accuracy that
   a. the reaction is endothermic.
   b. ΔS is positive.
   c. the reaction is spontaneous.
   d. the sign of ΔG is temperature dependent.
   e. $K > 1$

5. For the reaction

$$A + B \rightarrow C \qquad \Delta G° = -80\,kJ$$

   a. The reaction is nonspontaneous.
   b. When equilibrium is reached, the reaction mixture will be mostly C.
   c. At equilibrium, only A and B will be in the reaction mixture.
   d. $A + B \rightarrow C$ is exothermic.
   e. $C \rightarrow A + B$ is spontaneous.

B. Consider the following graph that represents the path of a reaction.

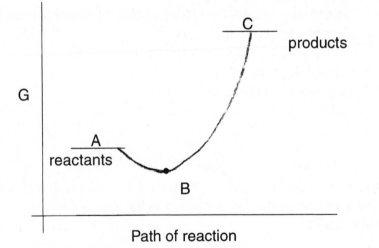

1. Fill in the blanks with $>$, $<$, or $=$.
   a. For the reaction between A and B: ΔG _____ 0 ; Q _____ K
   b. At B: ΔG _____ 0 ; Q _____ K
   c. Between B and C: ΔG _____ 0 ; Q _____ K

2. Is the reaction spontaneous?

C. Calculate ΔG at 25°C for the reaction

$$NH_3 \text{ (aq)} + H_2O \longrightarrow OH^- \text{ (aq)} + NH_4^+ \text{ (aq)}$$

when $[NH_3] = 0.10$, $[NH_4^+] = 0.10$, pH = 12.7 and K_a for $NH_4^+ = 5.6 \times 10^{-10}$.

D. For the reaction

$$2\,N_2H_4 \text{ (g)} + 2\,NO_2 \text{ (g)} \longrightarrow 3\,N_2 \text{ (g)} + 4\,H_2O \text{ (g)}$$

the following thermodynamic data is available:

	ΔH_f° (kJ/mol)	S° (kJ/mol-K)
N_2H_4	95.0	0.2385
NO_2	33.84	0.2404
N_2	0	0.1915
H_2O (g)	–241.8	0.1887

Calculate the partial pressure of steam at 100°C at the point in the reaction when P_{N_2} = 2.5 atm, $P_{N_2H_4}$ = P_{NO_2} = 1.0×10^{-3} atm and $\Delta G = -1264.6$ kJ.

Spontaneity of Reaction

Answers

Worksheet A

A. 1. a, b, c   2. d   3. d   4. b   5. d, e

B. 1. Spontaneity is T dependent, increasing as T increases.
   2. no
   3. yes
   4. $\approx$ 340 K ($\approx$ 67°C)
   5. 0.018

C. 56°C

D. 1. no   2. yes; above 2.2×10^4 K   3. no

Worksheet B

A. 1. False — <u>sometimes</u> spontaneous   2. True
   3. True   4. False — at <u>0 K</u>
   5. False — <u>sometimes</u> increases   6. True
   7. False — <u>increases</u>   8. False — <u>but not always</u> rapidly
   9. True   10. False — <u>is</u> a state ... path <u>independent</u>
  11. True   12. True
  13. True   14. False — <u>and yet do not have</u> the same
  15. False — <u>can</u> be spontaneous

B.

C. − 14.3 J/K   D. $\Delta S° = - 0.030$ kJ/K ; $\Delta H° = - 64.54$ kJ

Worksheet C

A. 1. b, c, d, e   2. a, e   3. c   4. b, c, e   5. b

B. 1. a. $<$, $<$ b. $=$, $=$ c. $>$, $>$
   2. no

C. 19.66 kJ

D. P_{H_2O} = 1.47 atm

Electrochemistry

Worksheet A

A. Circle the correct answer(s).

1. A substance that will reduce Ag^+ to Ag but will not reduce Ni^{2+} to Ni is

   a. Zn   b. Pb   c. Mg   d. Cd   e. Al

2. If all the species are in their standard states, which of the following is the strongest reducing agent?

   a. Br_2 b. Zn   c. Fe^{2+} d. Co^{3+} e. Mn

3. In a voltaic cell powered by a chemical reaction
   a. The reaction must be redox.   b. $E° > 0$
   c. $\Delta G° > 0$ d. $K = 0$
   e. Oxidation occurs at the anode.

4. For the cell

$$Fe \mid Fe^{2+} (0.1\ M) \parallel Cu^{2+} (0.1\ M) \mid Cu$$

   a. $E = E°$ b. $E < E°$ c. $E = 0$ d. $Q = 1$ $K = 1$

5. The following reaction takes place in a lead storage battery that is discharging.

$$PbO_2\,(s) + Pb\,(s) + 4\,H^+\,(aq) + 2\,SO_4^{2-}\,(aq) \rightarrow 2\,PbSO_4\,(s) + 2\,H_2O$$

   a. Pb is formed at the anode during discharge.
   b. PbO_2 is formed at the anode while the battery charges.
   c. pH increases as the battery discharges.
   d. The mass of Pb decreases while the battery is charging.
   e. The mass of $PbSO_4$ remains constant whether the battery is charging or discharging.

B. Consider the following description of a voltaic cell:

A 1.0 M solution of $Ni(NO_3)_2$ is placed in a beaker with a strip of nickel metal. A 1.0 M solution of $SnSO_4$ is placed in a second beaker with a strip of Sn metal. The two beakers are connected by a salt bridge and the two metal electrodes are linked by wire to a voltmeter.

1. What is the voltage generated by the cell under standard conditions?

2. Write a balanced (using smallest whole number coefficients) net ionic equation for the overall cell reaction.

3. Write the abbreviated form of the cell notation.

4. Draw the cell indicating
    a. the direction of electron flow.
    b. the anode.
    c. the cathode.

C. When 10.0 mL of 0.10 M Ce^{4+} is added to 25.0 mL of 0.20 M Fe^{2+}, the following reaction occurs:

$$Fe^{2+}(aq) + Ce^{4+}(aq) \rightarrow Fe^{3+}(aq) + Ce^{3+}(aq)$$

Given that

$$Ce^{4+}(aq) \rightarrow Ce^{3+}(aq) \qquad E^\circ_{red} = 1.610 \text{ V}$$

1. What is K at 25°C?

2. Find $[Ce^{4+}]$, $[Ce^{3+}]$, $[Fe^{2+}]$ and $[Fe^{3+}]$ at equilibrium.
    (*Hint:* If K is large, you can assume that the reaction goes to completion and the limiting reactant is used up.)

D. Acidified water is electrolyzed using copper electrodes. A steady current of 1.55 A is passed through the cell for 22 minutes and 35 seconds. The copper anode lost 0.692 g. No gas is evolved at the anode. At the cathode, hydrogen gas is collected over water at 27°C (vapor pressure H_2O = 26.74 mm Hg). The barometric pressure is 762.00 mm Hg.

1. What volume of hydrogen gas is collected?

2. What is the oxidation number of the copper ion formed?

Electrochemistry

Worksheet B

A. Determine whether the following statements are true or false. If the statement is false, rewrite it to make it true.

1. An atom changing oxidation number from –2 to –1 is being oxidized.

2. Reduction occurs at the anode.

3. An electrolytic cell generates electric current from a spontaneous redox reaction.

4. In both voltaic and electrolytic cells, anions migrate to the anode.

5. When $\Delta G° < 0$, the cell is voltaic and $E° < 0$.

6. Reactants at the top of the standard reduction potential table in your text are powerful oxidizing agents.

7. K can be determined from $E°$ as well as $\Delta G°$.

8. The voltage of a cell (E) depends on the amount of reactants, i.e., mass of solids, volume of liquids, concentration of ions, and pressure of gases.

9. The cell reaction
$$2\,Fe^{3+}\,(aq)\ +\ 2\,I^-\,(aq)\ \rightarrow\ 2\,Fe^{2+}\,(aq)\ +\ I_2\,(s)$$
will have the same $E°$ as
$$Fe^{3+}\,(aq)\ +\ I^-\,(aq)\ \rightarrow\ Fe^{2+}\,(aq)\ +\ \tfrac{1}{2}\,I_2\,(s)$$
but a different K.

10. In acidic solution, Fe (s) is a better reducing agent than Fe^{2+} (aq).

11. Mg can reduce Ag^+ to Ag but will not reduce Ni^{2+} to Ni.

12. Given that
$$A\ +\ \tfrac{1}{2}B\ \rightarrow\ 2\,C \qquad E°\ =\ +0.20\ V$$
then for the reaction
$$2\,A\ +\ B\ \rightarrow\ 4\,C \qquad E°$$
$E° = 2(0.2) = 0.40$ V.

13. If $E° = 1.00$ V for the forward reaction, then $E° = -1.00$ V for the reverse reaction.

14. Fluorine is the strongest reducing agent in the text's table of standard reduction potentials.

15. If K is very large (K > 10), we can assume that practically all the limiting reactant is used up.

B. Consider the following half-reactions:

$$2 H^+ (aq) + 2 e^- \rightarrow H_2 (g) \qquad\qquad E^\circ_{red} = 0.00\,V$$
$$O_2 (g) + 4 H^+ (aq) + 4 e^- \rightarrow 2 H_2O \qquad E^\circ_{red} = 1.229\,V$$

Show that the pH dependence of E for both reactions is the same even if it seems as if the second reaction would need a lower pH.

C. Given solutions of $FeSO_4$ (0.0500 M) and $CuSO_4$ (0.130 M), 25.0 g strips of Fe and Cu, and a KNO_3 salt bridge, construct a voltaic cell.

1. Diagram the cell.

2. Label the anode, cathode, and the direction of electron flow.

3. Write the abbreviated cell notation.

4. Calculate E°, ΔG°, and K for the cell.

5. Calculate E for the cell.

D. The same cell described in C is now attached to a cell battery and converted into an electrolytic cell. A current of 10.0 A is passed through the cell for two hours and 14 minutes. If the cell is 75% efficient, what are the masses of the metal strips after the current has passed through?

Electrochemistry

Worksheet C

A. Circle the correct answer(s).

1. The purpose of a salt bridge in a voltaic cell is
    a. to provide a pathway so that the reducing agent can reach the oxidizing agent.
    b. to provide a connection between the anode and the cathode electrodes so that the positive ions can gather around the cathode.
    c. to provide ions that will balance off the depletion or overabundance of positively or negatively charged species.
    d. to provide a connection between the anode, cathode, and voltmeter.
    e. to maintain the electron flow from anode to cathode.

2. In a voltaic cell
    a. an external source of current that acts as an "electron pump" is required.
    b. electrons flow toward the cathode.
    c. anions in the salt bridge travel away from the cathode.
    d. oxidation occurs at the cathode.
    e. $E° = E°_{red} - E°_{ox}$

3. The standard hydrogen electrode potential is 0.00 V at 25°C because
    a. the voltmeter records 0.00 V when the half reaction is set up.
    b. it has been measured accurately as 0.00 with respect to many electrodes.
    c. the hydrogen ion acquires electrons from a platinum electrode.
    d. H_2 (g) is the product of H^+ reduction and H_2 (g) is the naturally occurring form of hydrogen.
    e. it has been defined that way.
    f. hydrogen is the lightest element.

4. At standard conditions, a voltaic cell where $E° > 0$
    a. has $\Delta G° < 0$.   b. has $K > 1$.
    c. is an electrolytic cell.   d. is exothermic.
    e. has $\Delta S° < 0$.

5. Consider the standard reduction potential ($E°_{red}$) for gold
    $$Au^{3+} (aq) + 3e^- \rightarrow Au (s) \qquad E°_{red} = 1.498 \text{ V}$$
    From this information, one can surmise that
    a. gold metal is very difficult to oxidize, and thus to corrode.
    b. Au (s) is an excellent reducing agent.
    c. Au^{3+} (aq) is an excellent reducing agent.
    d. $2 Au^{3+} (aq) + 3 H_2 (g) \rightarrow 6 H^+ (aq) + 2 Au (s) \qquad \Delta G° < 0$
    e. gold is expensive.

B. Currently, E_{red}° values for half cells are measured by comparison with

$$2\,H^+\,(aq) + 2\,e^- \rightarrow H_2\,(g) \qquad E_{red}^{\circ} = 0.00\ V$$

Suppose that the half reaction

$$AuCl_4^-\,(aq) + 3\,e^- \rightarrow Au\,(s) + 4\,Cl^-\,(aq) \qquad E_{red}^{\circ} = 1.00V$$

is designated as the new standard where E_{red}° is zero.

Answer the following questions with reference to the current standard reduction potential table in your text. Use **I** to indicate an increase, **D** to indicate a decrease, and **S** to indicate no change.

_____ 1. E° for net cell reactions where $E^{\circ} = E_{red}^{\circ} + E_{ox}^{\circ}$?

_____ 2. The number of half cells that would have $E_{red}^{\circ} < 0$?

_____ 3. E_{ox}° for Li (s), the strongest reducing agent?

_____ 4. E_{red}° for $Co^{3+}\,(aq) + e^- \rightarrow Co^{2+}\,(aq)$?

_____ 5. Effect of the sign of E° on spontaneity?

C. Consider the following standard reduction potentials:

$$Tl^+\,(aq) + e^- \rightarrow Tl\,(s) \qquad\qquad E_{red}^{\circ} = -0.34\ V$$
$$Tl^{3+}\,(aq) + 3\,e^- \rightarrow Tl\,(s) \qquad\qquad E_{red}^{\circ} = 0.74\ V$$
$$Tl^{3+}\,(aq) + 2\,e^- \rightarrow Tl^+\,(aq) \qquad\quad E_{red}^{\circ} = 1.28\ V$$

and the following abbreviated cell notations

(1) $Tl\mid Tl^+\parallel Tl^{3+}\mid Tl^+\mid Pt$
(2) $Tl\mid Tl^{3+}\parallel Tl^{3+}\mid Tl^+\mid Pt$
(3) $Tl\mid Tl^+\parallel Tl^{3+}\mid Tl$

1. Write the overall equation for each cell.
2. Calculate E° for each cell.
3. Calculate ΔG° for each overall equation.
4. Comment on whether ΔG° and/or E° are state properties.
    (*Hint:* A state property is path independent.)

D. From the half cell equations listed below, determine the K_{sp} of Hg_2I_2.

$$Hg_2^{2+}\,(aq) + 2\,e^- \rightarrow 2\,Hg\,(\ell) \qquad\qquad E_{red}^{\circ} = 0.789\ V$$
$$Hg_2I_2\,(s) + 2\,e^- \rightarrow 2\,Hg\,(\ell) + 2\,I^-\,(aq) \qquad E_{red}^{\circ} = -0.041\ V$$
$$I_2\,(s) + 2\,e^- \rightarrow 2\,I^-\,(aq) \qquad\qquad\qquad E_{red}^{\circ} = 0.530\ V$$

E. Consider the following voltaic cell:

$$Fe\mid Fe^{2+}\,(2.0 \times 10^{-5}\ M)\parallel H^+\,(3.4\ M)\mid H_2\,(1.00\ atm)\mid Pt$$

Find
1. E°.
2. E at 25°C.
3. E at equilibrium.
4. K at 25°C.

Electrochemistry

Answers

Worksheet A

A. 1. b   2. e   3. a, b, e   4. a, d   5. b, c

B. 1. 0.095 V
2. $Ni(s) + Sn^{2+}(aq) \rightarrow Sn(s) + Ni^{2+}(aq)$
3. $Ni \mid Ni^{2+} \parallel Sn^{2+} \mid Sn$
4.

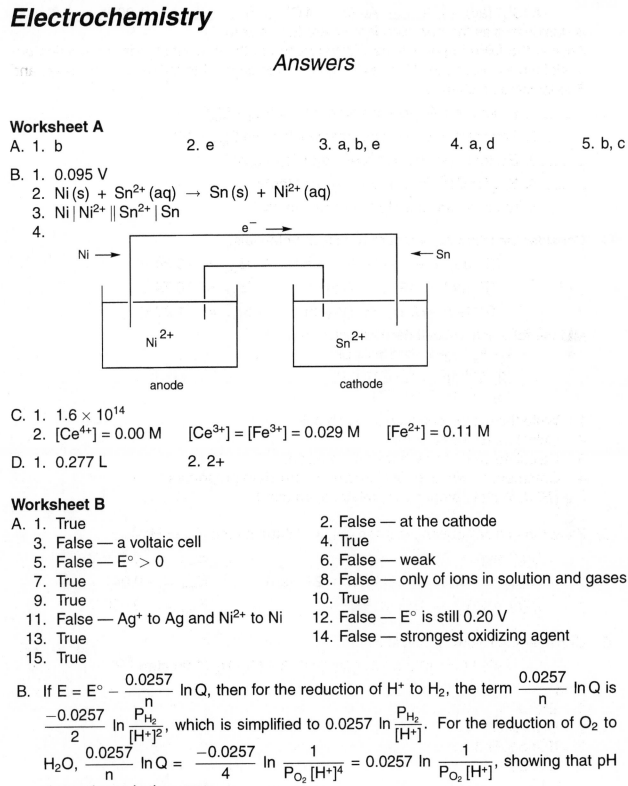

C. 1. 1.6×10^{14}
2. $[Ce^{4+}] = 0.00$ M $[Ce^{3+}] = [Fe^{3+}] = 0.029$ M $[Fe^{2+}] = 0.11$ M

D. 1. 0.277 L   2. 2+

Worksheet B

A. 1. True   2. False — at the cathode
    3. False — a voltaic cell   4. True
    5. False — $E° > 0$ 6. False — weak
    7. True   8. False — only of ions in solution and gases
    9. True   10. True
  11. False — Ag^+ to Ag and Ni^{2+} to Ni 12. False — $E°$ is still 0.20 V
  13. True   14. False — strongest oxidizing agent
  15. True

B. If $E = E° - \dfrac{0.0257}{n} \ln Q$, then for the reduction of H^+ to H_2, the term $\dfrac{0.0257}{n} \ln Q$ is $\dfrac{-0.0257}{2} \ln \dfrac{P_{H_2}}{[H^+]^2}$, which is simplified to $0.0257 \ln \dfrac{P_{H_2}}{[H^+]}$. For the reduction of O_2 to H_2O, $\dfrac{0.0257}{n} \ln Q = \dfrac{-0.0257}{4} \ln \dfrac{1}{P_{O_2}[H^+]^4} = 0.0257 \ln \dfrac{1}{P_{O_2}[H^+]}$, showing that pH dependence is the same.

C. 1–2

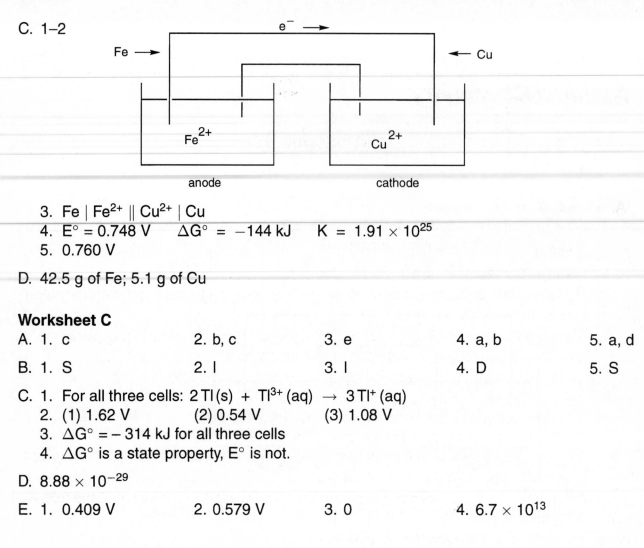

3. Fe | Fe^{2+} ‖ Cu^{2+} | Cu
4. E° = 0.748 V   ΔG° = −144 kJ   K = 1.91 × 10^{25}
5. 0.760 V

D. 42.5 g of Fe; 5.1 g of Cu

Worksheet C

A. 1. c   2. b, c   3. e   4. a, b   5. a, d

B. 1. S   2. I   3. I   4. D   5. S

C. 1. For all three cells: 2 Tl (s) + Tl^{3+} (aq) → 3 Tl$^+$ (aq)
   2. (1) 1.62 V   (2) 0.54 V   (3) 1.08 V
   3. ΔG° = − 314 kJ for all three cells
   4. ΔG° is a state property, E° is not.

D. 8.88 × 10^{-29}

E. 1. 0.409 V   2. 0.579 V   3. 0   4. 6.7 × 10^{13}

Nuclear Chemistry

Worksheet A

A. Circle the correct answer(s).

1. For stability, the optimal neutron–proton ratio in a nuclide is one. Which of the following nuclides is the least stable?

a. $_2^4$ He   b. $_6^{13}$ C   c. $_{53}^{127}$ I   d. $_{82}^{201}$ Pb   e. $_{85}^{210}$ At

2. $_{20}^{47}$ Ca is an unstable isotope. In order to approach the optimal neutron–proton ratio, the isotope decays by

a. emitting α–particles.   b. emitting β–particles.   c. emitting positrons.
d. emitting γ–rays.   e. K–electron capture.

3. Neutron bombardment of an isotope yields as the first product
   a. the same element of greater mass.
   b. an element of the same mass but lower atomic number.
   c. an element of the same mass but higher atomic number.
   d. an unstable nuclide made up of the same mass and the same atomic number as the isotope.

4. Rutherford's alpha particle scattering experiment established which of the following characteristics?
   a. the charge to mass ratio of protons
   b. the approximate diameter of gold atoms
   c. the concentration of atomic masses in the nucleus
   d. the uniform distribution of electrical charge throughout the atom

5. The atomic mass of uranium is 238.03 amu, and the two most abundant isotopes are $_{92}^{238}$ U and $_{92}^{235}$ U.
   a. U–238 has more protons than U–235.
   b. U–238 has more neutrons than U–235.
   c. Electrically neutral atoms of U–235 and U–238 have the same number of electrons.
   d. U–238 is more abundant than U–235.

6. The atomic mass of a tritium atom, $_1^3 H$, is listed in a reference book as 3.01605 amu. This is less than the sum of the masses of the proton, electron, and two neutrons that make up the atom. This is best explained by which statement(s)?
   a. The masses of the protons, neutrons, and electrons were probably taken from a different reference book.
   b. The tritium atom is held together by energy equivalent to the difference in mass.
   c. Energy must be added to a proton plus an electron plus 2 neutrons to form a tritium atom.
   d. Tritium is a β–emitter.

B. Write and balance the nuclear reactions for the following processes.
   1. The β decay of K–47.
   2. Positron emission by As–72
   3. The α–decay of Th–226
   4. K–electron capture by Er–160
   5. α–particle bombardment of N–14 producing a proton and another nuclide.

C. Calculate the binding energy of Si–34 (33.9764 amu). The masses of a proton and a neutron are 1.00728 and 1.00867 amu, respectively.

D. A series of experiments require Co–60 with an activity of at least 4.00 mCi. Co–60 is a β–emitter with a half life of 5.26 years.
   1. If the Co–60 bought by the lab has an activity of 7.00 mCi, what is the shelf life of Co–60 for this particular series of experiments?

   2. What percent of the activity remains after 2.0 years?

E. Plutonium–239 is an α–emitter and decays according to the following nuclear reaction:
$$_{94}^{239} Pu \rightarrow _2^4 He + _{92}^{235} U$$
   1. Calculate the energy obtained from the decay of two hundred milligrams of Pu–239.

   2. Suppose that 20.0% of the energy obtained from the decay of Pu–239 is converted to electrical energy and designated as w_{max}. How many volts from a Nicad battery ($Cd \mid Cd^{2+} \parallel Ni^{2+} \mid Ni$) is that energy equivalent to?

Nuclear Chemistry

Worksheet B

A. Circle the true statement(s).

1. A positron has
   a. the same mass as an electron.
   b. the same mass as an α–particle.
   c. the same mass as a proton.
   d. a charge of +1.
   e. zero charge.

2. The rate of radioactive decay (activity) is
   a. directly proportional to the amount of radioactive isotope present.
   b. inversely proportional to the rate constant.
   c. always first order.
   d. increases as the amount of radioactive isotope decreases.

3. The half life of a radioactive compound
   a. depends on the amount of radioactive isotope present.
   b. is inversely proportional to the rate constant.
   c. is always 0.693.
   d. increases when the mass of radioactive isotope increases.

4. What particle is deflected most when moving between two oppositely charged plates?
   a. positron
   b. α–particle
   c. γ–rays
   d. β–particle
   e. Both the positron and β–particle are deflected but in opposite directions.

5. In a nuclear reaction
   a. the number of protons is conserved but not the number of neutrons.
   b. a proton is the same as H–1.
   c. there is always a loss of mass.
   d. $\Delta E > 0$ means that the mass of products is larger than the mass of reactants.
   e. when $\Delta m < 0$, the reaction is spontaneous.

B. Write and balance nuclear reactions for the following processes.
    1. Thorium–234 decays to protactinium (Pa) – 234.
    2. The bombardment of uranium-235 with a neutron yielding xenon–135, two neutrons, and another nuclide.
    3. The bombardment of nitrogen–14 with a particle yielding oxygen–17 and a proton.
    4. The K–electron capture of aluminum–25.
    5. The α–particle bombardment of copper–63 yielding a neutron and another nuclide.

C. Fill in the following table

Event	Change in atomic number	Change in number of neutrons	Change in mass number
α–decay	_____	_____	_____
β–emission	_____	_____	_____
positron emission	_____	_____	_____
K–electron capture	_____	_____	_____
γ–decay	_____	_____	_____

D. Thorium has a half life of 1.4×10^{10} years.
    1. What is the activity (rate of decay) of a one nanogram sample (1.0×10^{-9} g) of Th–232 in atoms/year and Ci?

    2. From the activity obtained in (1), how long (in days) will it take for one atom to decay?

    3. How long will it take for 1% of a sample of Th–232 to decay?

E. Actinium–227 has a half life of 8×10^3 days, decaying by α–emission. Suppose that the helium gas originating from the alpha particles were collected. What volume of helium at 25°C and 748 mm Hg could be obtained from 1.00 g of Ac–227 after 100 years of decay?

Nuclear Chemistry

Worksheet C

A. Fill in each blank with the correct answer chosen from the list on the right. An answer can be used only once. There are more choices than correct answers.

_____ 1. The emission of a helium nucleus by a radio-active nucleus is called __(1)__ .

_____ 2. If 1 mg of a radioactive sample is left after a 4 mg sample is allowed to decay for 60 min, the half life for the decay is __(2)__ .

_____ 3. The emission of an electron from the nucleus is called __(3)__ .

_____ 4. A nuclear reaction in which an inner shell electron is used to convert a proton to a neutron is called __(4)__ .

_____ 5. A heavy isotope splitting into 2 atoms of intermediate mass and several neutrons is called __(5)__ .

_____ 6. A reaction or series of reaction steps that initiates repetition of itself is called __(6)__ .

_____ 7. 0.693/k is equal to __(7)__ .

_____ 8. The combination of 2 light nuclei to give a heavier nucleus is called __(8)__ .

_____ 9. Reactions that result in changes in the atomic number, mass number, or energy of the nucleus is called __(9)__ .

_____10. The unit for activity is __(10)__ .

a. α–decay
b. β–decay
c. chain reaction
d. 30 min
e. half life
f. K–electron capture
g. fission
h. fusion
i. nuclear reaction
j. 15 min
k. disintegrations
l. Curie
m. rem

B. The following data is obtained for a radioactive isotope. Plot the data and determine the half life of the isotope.

Time (hour)	0.00	0.50	1.00	1.50	2.00	2.50
Activity (disintegrations/hr)	14,472	13,095	11,731	10,615	9,605	8,504

C. The cleavage of ATP (adenosine triphosphate) to ADP (adenosine diphosphate) and
    H_3PO_4 may be written as follows:

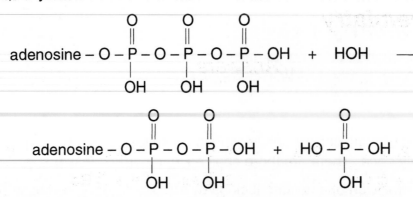

It is interesting to determine which bond (the P–O bond marked **a** or the O–P bond
marked **b**) is cleaved by hydrolysis (reaction with water).

1. Outline an experiment (using radioactivity) that can be used to determine where
    the cleavage results.

2. Describe the results that would lead you to conclude that cleavage results at **a** .

3. Describe the results that would lead you to conclude that cleavage results at **b** .

Results show that the cleavage occurs at **b** .

D. Consider a radioactive isotope in which 0.0020 mol undergoes 5.5×10^{13} disintegra-
    tions (atoms)/s.
    1. What is its activity in curies?
    2. What is its half life in days?
    3. What percent of the original sample disintegrates in 57 days?

E. The half life of Tc–99 is 6.0 hr. The delivery of a sample of Tc–99 from the reactor
    to a lab of a certain hospital takes one hour and 22 minutes. What is the minimum
    amount of Tc–99 that can be shipped in order for the hospital to receive 10.0 mg of
    Tc–99?

Nuclear Chemistry

Answers

Worksheet A

A. 1. e   2. b   3. a   4. c   5. b, c, d   6. b

B. 1. $^{47}_{19}K \rightarrow {}^{0}_{-1}e + {}^{47}_{20}Ca$ 2. $^{72}_{33}As \rightarrow {}^{0}_{1}e + {}^{72}_{32}Ge$

   3. $^{226}_{90}Th \rightarrow {}^{4}_{2}He + {}^{222}_{88}Ra$ 4. $^{160}_{68}Er + {}^{0}_{-1}e \rightarrow {}^{160}_{67}Ho$

   5. $^{14}_{7}N + {}^{4}_{2}He \rightarrow {}^{1}_{1}H + {}^{17}_{8}O$

C. 2.69×10^{10} kJ/mol

D. 1. 4.25 years   2. 77%

E. 1. 4.3×10^{5} kJ   2. 448 V

Worksheet B

A. 1. a, d   2. a, c   3. b   4. a, d, e   5. b, d, e

B. 1. $^{234}_{90}Th \rightarrow {}^{0}_{-1}e + {}^{234}_{91}Pa$ 2. $^{235}_{92}U + {}^{1}_{0}n \rightarrow 2\,{}^{1}_{0}n + {}^{99}_{38}Sr + {}^{135}_{54}Xe$

   3. $^{14}_{7}N + {}^{4}_{2}He \rightarrow {}^{1}_{1}H + {}^{17}_{8}O$ 4. $^{25}_{13}Al + {}^{0}_{-1}e \rightarrow {}^{25}_{12}Mg$

   5. $^{63}_{29}Cu + {}^{4}_{2}He \rightarrow {}^{1}_{0}n + {}^{66}_{31}Ga$

C. α–decay: –2, –2, –4 β–emission: + 1, –1, 0   positron emission: –1, +1, 0
   K–electron capture: –1, +1, 0 γ–decay: 0, 0, 0

D. 1. 1.3×10^{2} atoms/yr; 1.1×10^{-16} Ci
   2. 2.9 days
   3. 2.0×10^{8} yr

E. 104 mL

Worksheet C

A. 1. a   2. d   3. b   4. f   5. g
   6. c   7. e   8. h   9. i   10. l

B.

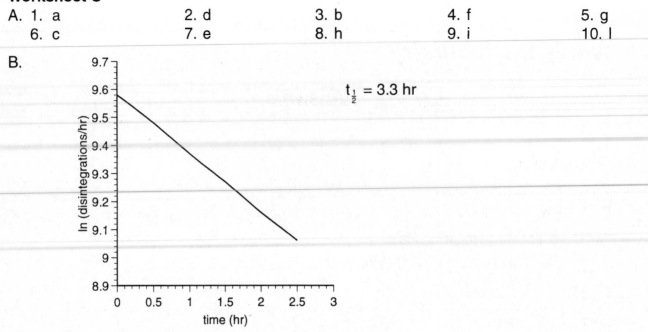

C. 1. Prepare water where the oxygen atom is radioactive. Hydrolyze ATP with it and separate the products (ADP and H_3PO_4) from each other. Check ADP and H_3PO_4 for the presence of the radioactive isotope of oxygen.

   2. If ADP has the radioactive isotope of oxygen, ATP cleaves at **a** .

   3. If H_3PO_4 has the radioactive isotope of oxygen, ATP cleaves at **b** .

D. 1. 1.5×10^3 Ci   2. 174 days   3. 20%

E. 12 mg